Energy and Transport Processes

Core Programming Topic at the 2012 AIChE Annual Meeting

Pittsburgh, Pennsylvania, USA
28 October - 2 November 2012

ISBN: 978-1-62276-725-0

Printed from e-media with permission by:

Curran Associates, Inc.
57 Morehouse Lane
Red Hook, NY 12571

Some format issues inherent in the e-media version may also appear in this print version.

Printed by Curran Associates, Inc. (2013)

For permission requests, please contact AIChE
at the address below.

AIChE
3 Park Avenue
New York, NY 10016-5991

Phone: (203) 702-7660
Fax: (203) 775-5177

www.aiche.org

Additional copies of this publication are available from:

Curran Associates, Inc.
57 Morehouse Lane
Red Hook, NY 12571 USA
Phone: 845-758-0400
Fax: 845-758-2634
Email: curran@proceedings.com
Web: www.proceedings.com

TABLE OF CONTENTS

(29a) Salt Effect On the Rheology of Hydrate-Forming Emulsions

Genti Zylyftari[1,2], Jae W. Lee[1] and Jeffrey F. Morris[1,3], (1)Chemical Engineering, City College of City University of New York, New York, NY, (2)Levich Institute, City College of New York, New York, NY, (3)Levich Institute, City College of City University of New York, New York, NY

Clathrate hydrate formation within pipelines is a major problem in the oil and gas industry. It causes flow interruption and significant equipment damage. Hydrate, a crystalline compound, forms when the hydrogen-bonded water cage is stabilized by a light hydrocarbon molecule at low temperatures and high pressures, conditions predominantly found in subsea pipelines. An efficient and cost effective mitigation of plug formation requires a mechanical understanding of the process involved in hydrate formation. Since offshore drilling occurs under the seafloor, the salinity of the aqueous phase in the aquifers is an important factor to be considered for the formation of hydrate. Salt is a thermodynamic inhibitor of hydrate. We study the effect of salinity on a 40% (*v/v*) aqueous phase density-matched cyclopentane hydrate-forming emulsion through a combination of experimental techniques using a micro-differential scanning calorimeter (μ-DSC) and rheometer. Cyclopentane hydrate-brine system liquidus line is plotted using calorimetric data. Equilibrium temperature and thermodynamical conversion of water to hydrate particles in the emulsion are obtained from the hydrate-brine phase diagram. Rheological properties - viscosity critical time, viscosity evolution time and final viscosity - of the hydrate-forming emulsion are extracted from the rheological experiments. A correlation of the rheological properties of the hydrate-forming emulsion to the thermodynamic driving forces is obtained. Viscosity critical time and evolution time decrease with higher subcooling. Final viscosity dependence on the thermodynamical water to hydrate conversion shows that both aqueous phase and hydrate content are intrinsically important to the strength of formed network like structures. Maximum viscosity is attained when the thermodynamic water to hydrate conversion is in the 70-80% range. The hydrate-forming emulsion final structure exhibits shear thinning effects, which are fitted to the Ostwald-de Waele model.

(29b) Modeling Hydrate Formation and Aggregation in Oil-Dominated Systems

Zachary M. Aman, Luis Zerpa, Giovanny Grasso, Kyle Springer, E. Dendy Sloan, Amadeu K. Sum and Carolyn A. Koh, Chemical & Biological Engineering, Colorado School of Mines, Golden, CO

Gas hydrates are ice-like solids, where hydrogen-bonded water molecules form cages surround light hydrocarbon gases (e.g., methane), typically at high pressure and low temperature. Hydrate formation can lead to operational and safety hazards in oil and gas production flowlines, Thermodynamic inhibition of gas hydrates is typically used to prevent the formation of these solids, but this method becomes prohibitively expensive as water production increases over the lifetime of the field. In recent years, the flow assurance aspect of hydrates has shifted from prevention to risk management, which requires a greater understanding of the formation and accumulation processes of hydrates in flowlines. One component to assess the impact of hydrates on flowlines is the impact of hydrates on flow properties. Here we present a model to capture transient oil-phase viscosification and pressure drop during hydrate formation and particle aggregation. The model is based on dynamic aggregation of particles, derived from cohesion force measurements between hydrate particles. This model captures transient effects of combined hydrate formation and aggregation. We also present a comparison of the model to several experimental datasets for hydrate formation and pressure drop collected at a large-scale industrial flowloop. The experiments were collected over a variety of water cuts (15-90 vol%), liquid loadings (50-90 vol%), and flow rates (1-3 m/s). Compared to previous models based on uniform aggregate size, the results show better predictions of the transient formation and aggregation behavior observed in flowloop experiments.

(29c) Investigating the Role of Wax Gelation in Wax Deposition and Growth in Subsea Pipelines

Yingda Lu, Chemical Engineering, University of Michigan, Ann Arbor, MI and Zhenyu Huang, Multiphase Solutions Kenny, Incorporated

Abstract

Wax deposition during the transportation of crude oil in the subsea pipelines has caused significant flow assurance problems in the petroleum industry. One practical but costly way used to remediate wax deposition is pigging, which involves sending inspection gauge or 'pig' along the length of the pipeline to scrape off the wax. In order to arrive at an optimal pigging frequency, the growth rate of wax deposition needs to be accurately determined. Wax deposition models can serve as predictive tools for estimating the growth rate of wax deposition in subsea pipelines.

Most of the current wax deposition models were developed based on the mechanism of molecular diffusion, which was developed by *Probjot et. al.* (AIChE J. **2000**, 46, 1059–1074) and was highly regarded among wax deposition studies. However, our recent analyses have shown that another mechanism might play a role in addition to molecular diffusion contributing to the deposit growth, which is overlooked in most previous studies: at low oil flow rates (i.e. low shear rates), it is expected that not only molecular deposition takes place at the oil-deposit interface, but the wax molecules at the boundary layer might crystallize to form a cross-linked network and help to further grow the deposit. This gelation phenomenon is unlikely to occur at high oil flow rates because the high shear stress tends to break up the crystal network.

The carbon number distribution of the deposit obtained from gas chromatography is a significant indicator of the role of gelation in the wax deposit growth. As the solids in the deposit consist of mostly the heavy components, a higher mass fraction of heavy components in the carbon distribution indicates a larger amount of solid formed. The difference in the carbon number distribution between the deposit and the oil (above the WAT) is that the heavy components are enriched in the deposit due to the diffusion of wax molecules (heavy alkanes). As the diffusion-driven formation represents the continuous radial transport of wax molecules depositing on the interface and diffusing into the gel, its resulting deposit is expected to contain more solids than a gelation-driven deposit. Consequently, the deposit formed mainly by diffusion-driven deposition is expected to have higher mass fractions of heavy alkanes than that formed by gelation. The overall relationship between the flow rate and carbon number distribution in GC is shown in Table 1.

Table 1: Influence of decreasing flow rate on the deposit gas chromatography

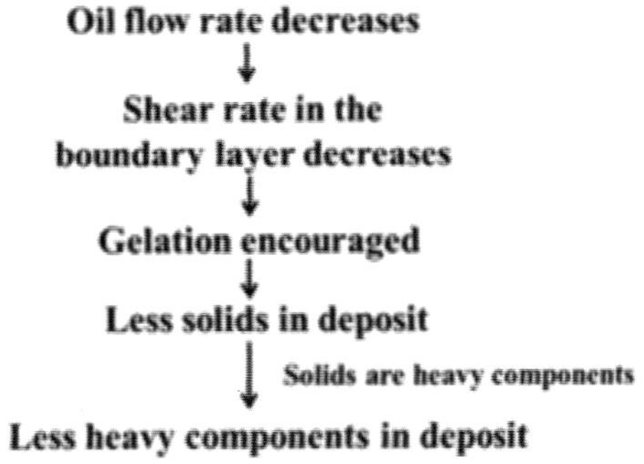

The GC analyses of the deposits obtained from flow-loop experiments have been carried out and the results are shown in Figure 1.

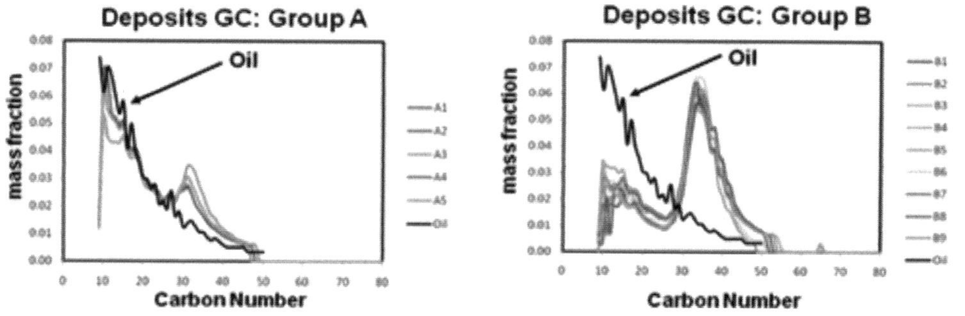

Figure 1: GC analyses of experiments

It is clearly seen that the carbon number distribution of the experiments can be divided by two groups of experiments that are formed by two drastically different mechanisms. The experiments carried out at relatively low flow rates and low wall temperatures (Group A) show very small amounts of heavy components (low solid fraction) in the deposit, indicating that the deposit is gelation-driven. The experiments at relatively high flow rates and high wall temperatures (Group B) show a great contents of heavy components (high solid fraction), which implies that the deposit is deposition-driven.

Therefore, a fundamental investigation of the role of wax gelation in wax deposit growth will be carried. The work will start with several rheometric experiments to investigate the effect of shear history and cooling rate on gelation temperature. The goal is to quantitatively describe the gelation phenomena and develop a theoretical model for gelation kinetics. The finally purpose would be to incorporate the gelation model into the current wax deposition models to improve the prediction quality.

(29d) Gelation and Rheological Behavior of Wax-Oil Systems

Yansong Zhao[1], Lalit Kumar[2], Kristofer Paso[2] and Johan Sjöblom[2], (1)Department of Chemical Engineering, Norwegian University of Science and Technology, Trondheim, Norway, (2)Department of Chemical Engineering, Norwegian University of Science and Technology, Trondheim, Norway

Gelation and Rheological Behavior of Wax-Oil Systems

Yansong Zhao*, Lalit Kumar, Kristofer Paso, Johan Sjöblom

Ugelstad Laboratory, Department of Chemical Engineering, Norwegian University of Science and Technology (NTNU), Trondheim 7491, Norway

*Corresponding author. E-mail address: yansong.zhao@ntnu.no (Yansong Zhao)

Tel: +47-73550339

Abstract

Crude oil is a complex mixture of hydrocarbons, containing paraffins, aromatics, naphthenes, asphaltenes, and resins.[1] Gelation and rheological behavior is significant in waxy crude oil transportation strategies. Shut-in and restart of transport pipelines often result in reducing the temperature of crude oil fluid below the pour point temperature (PPT). Subsequently, paraffin wax gelation at the low temperature conditions leads to the strong solid-liquid gels formation.[2]The mechanisms of waxy crude oil gelation and formed gel breakage are important for waxy crude oil transportation.

Therefore, gelation and rheological behavior of wax-oil systems (model wax-oils and waxy crude oils) is investigated in this work. Differential scanning calorimetry, rheometry, microscopy and X-ray diffraction are the common methods used for wax-oil systems gelation and rheology investigation. Subsequently, 1 w/w% to 20 w/w% macro-crystalline wax in dodecane, 1 w/w% to 20 w/w% micro-crystalline wax in dodecane, 1 w/w% to 30 w/w% xylene in model wax-oil, 0.01 w/w% to 0.08 w/w% asphaltene in model wax-oil and 5 w/w% macro-crystalline wax in Primol 352 are prepared. Furthermore, effect of shear history, thermal history, wax content, wax type, xylene, asphaltene and additives on model wax-oil gelation and gel breakage process is investigated. It is shown that the gel strength decreases when shear stress is applied to the sample during the gel formation process. Larger crystals can be formed at relative lower cooling rates, which lead to stronger gels. Yield stress decreases with increasing asphaltene content ranging from 0.01 w/w% to 0.08 w/w% in model wax-oil used in this work. XRD results show that crystal lattice still forms even with the addition of a high-efficiency wax inhibitor.[2]Novel gelation and gel breakage models are established based on model wax-oil experiments and previous modeling results. Moreover, waxy crude oils have been used to check the novel gelation and gel breakage models. It is shown that the novel models are applicable for the waxy crude oils, which is useful for the shut-in and restart process of waxy crude oil transportation prediction.

(1) Venkatesan, R.; Östlund, J. A.; Chawla, H.; Wattana, P.; Nydén, M.; Fogler, H. S. *Energy Fuels* **2003,** *17*(6), 1630-1640.

(2) Zhao, Y. S.; Kumar, L.; Paso, K.; Ali, H.; Safieva, J.; Sjöblom, J. *Ind. Eng. Chem. Res.* (*Revised*)

(29e) Naphthenic Acids in Oils with Different Contents of Asphaltenes

Fenglou Zou, DBR Technology Center, Edmonton, AB, Canada, Huang Zeng, Schlumberger, DBR Technololy Center, Edmonton, AB, Canada and Simon Andersen, DBR Technology Center, Schlumberger, Edmonton, AB, Canada

Naphthenic acids in oils with different contents of asphaltene

Fenglou Zou, Huang Zeng, Simon I. Andersen

Schlumberger, DBR Technology Center, T6N 1M9 Edmonton, Canada

Carboxylic acids (RCOOH), also known as naphthenic acid in the oil field industry, is characterized by the presence of at least one carboxyl group (-COOH). They are widespread compounds, occurring in sediments and petroleums. Due to the structural characteristics, naphthenic acids are considered as natural surfactants in crude oil and believed to play a part in oil viscosity, emulsion stability, reservoir wettability, and enhanced oil recovery by alkaline flooding as well as corrosion. Therefore the knowledge of acid content and acid structure may have a significant practical application potential. The acid content in crude oils varies. One may speculate on the relationship between crude API, asphaltene content and the content of acids. As such there is currently no indication in the open literature of a straight relation.

Two oils (A and B) containing different asphaltene contents were extracted with either ion exchange resins (IER) or water. A (4% asphaltene), has higher acid concentration than B (14% asphaltene). Obviously higher asphaltene content does not mean necessarily higher acid concentration. Figure 1 compares the speciation of acid composition of the two oils using 2D GC/MS after methylation. B contains more diacids and straight chain acids but less aromatic acids and cyclic acids. The water extraction was found to yield completely different acid distributions when compared to the IER extracted. Both mono and diacidic compounds were detected. The diacids with two carboxylic acid groups could have structure building potential leading to increased non-Newtonian behaviour. Straight chain acids are known to act as surfactants at interfaces. More diacids were found in B compared to A implying that higher asphaltene content could indicate more complexity in the acids. Data from a range of oils will be presented.

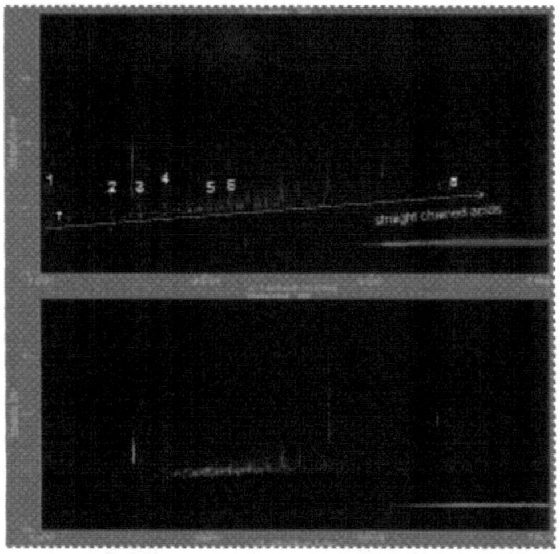

Fig. 1: Comparison of ion (m/z = 59) chromatograms of oil B 14% asp. (top) with oil A 4% asp. (bottom). 1: ethandioic acid, dimethyl ester, 2: butanedioic acid, dimethyl ester, 3: pentanedioic acid, dimethyl ester, 4: hexanedioic acid, dimethyl ester, 5: octanedioic acid, dimethylester, 6:nonanoic acid, dimethyl ester 7: C_4H9COOH, 8: $C_{28}H_{57}COOH$

(29f) Insight Into the Physical Asphaltene Precipitation Mechanism by Small-Angle Neutron Scattering

Michael P. Hoepfner, Claudio Vilas Boas Favero and H. Scott Fogler, Chemical Engineering, The University of Michigan, Ann Arbor, MI

The discovery of slow asphaltene precipitation kinetics has provided a new avenue to probe asphaltene behavior to discover both the physical precipitation mechanism and the factors that control the process. Results published by Maqbool, *et al.* revealed that even though a mixture of oil and precipitant appeared stable, if monitored for up to several months, asphaltene precipitation could be detected [1]. Small-angle neutron scattering (SANS) results presented at Petrophase XII, revealed that the slow precipitation kinetics are the result of detection limitations associated with standard laboratory practices [2]. When SANS was used on similar systems, the instability could be detected nearly instantly due to the sensitivity of small-angle scattering to nanometer length-scale changes.

A new model has been developed to analyze SANS results of slowly precipitating asphaltene systems that not only fits the scattering results, but also predicts the fraction of insoluble and precipitating asphaltenes within reasonable accuracy. One of the inherent major limitations of small-angle scattering analysis is a lack of validation for the models selected to fit the results. Typically the only evidence for correctness of a particular scattering model is the quality of the fit. However, in this approach we have compared the predictions generated from analysis of the scattering results to the solubility of asphaltenes after we allowed the slow precipitation kinetics to resolve.

According to the new model, the physical mechanism for asphaltene precipitation is that asphaltene fractal aggregates or "clusters" grow in size once a precipitant is added. Results suggest that when the size of the clusters reach a critical value, a fraction precipitate and a geometric population balance has been previously utilized to show that this final stage of precipitation is likely a reaction-limited aggregation process [3]. Based on the model, the scattering observed is due primarily to the interaction of asphaltenes and the physical structure of the clusters. The asphaltenes not precipitating remain as stable clusters and at a size smaller than in the original oil. Soluble asphaltenes at a specific precipitant concentration will be of a more stable fraction and will have weaker interactions. In conclusion, we believe that the success of the model to predict asphaltene solubility strongly supports the described precipitation mechanism and represents an important step toward deciphering the complex behavior of asphaltenes.

[1] T. Maqbool, A. T. Balgoa, and H. S. Fogler, *Energy & Fuels*, 23, 3681-3686 (2009).

[2] M. P. Hoepfner, C. V. B. Favero and H. S. Fogler, Oral Presentation, *Petrophase XII*, London, UK, 10-14 Jul. (2011).

[3] T. Maqbool, S. Raha, M. P. Hoepfner, and H. S. Fogler, *Energy & Fuels*, 25, 1585-1596 (2011).

(29g) Properties Controlling Aggregation and Precipitation Rates of Asphaltenes

Nasim Haji Akbari Balou and H. Scott Fogler, Chemical Engineering, The University of Michigan, Ann Arbor, MI

Properties Controlling Aggregation and Precipitation Rate of Asphaltenes

Nasim Haji Akbari Balou[1] and H. Scott Fogler[1]

[1]The University of Michigan, Department of Chemical Engineering, Ann Arbor, MI

Asphaltenes are the heaviest and most polar fraction of crude oil and can precipitate due to change in temperature, pressure and composition. Their precipitation and deposition can lead to serious economical impacts in the oil industry. For years it was believed that instability of asphaltenes can be detected shortly after changing the operational conditions. However, recent studies in our group demonstrated that asphaltene-precipitant mixtures can be unstable even at very low precipitant concentrations and it can take weeks and sometimes even months to detect such instabilities [1, 2]. We have observed that this kinetic behavior is universal among different crude oils. However, the precipitation rates strongly depend on the properties of the crude oil. Our goal in this study is to identify the properties that are responsible for differing aggregation and precipitation kinetics. Due to the complexity of crude oil, model mixtures are used as alternatives to measure the precipitation rates for different solvents and asphaltenes.

A new analysis approach has been developed to estimate the precipitation rates by accounting for the properties governing the aggregation kinetics such as viscosity and solubility parameter. The results show that precipitation rates strongly depend on the solvent used for asphaltenes stabilization. However, we show that under certain assumptions all the differences for different types of asphaltenes and solvents can be well explained, and their precipitation rates collapse onto a single curve. The results also demonstrate that not only the solvent, but also the asphaltene content of the solution can play an important role in controlling the aggregation rate. We have observed that solutions with lower asphaltene content show a higher tendency for destabilization and eventual precipitation. These findings will lead to a better understanding of the properties that govern the destabilization and growth processes of asphaltenes.

[1] T.Maqbool, A.T. Balgoa, and H. S. Fogler, " Revisiting Asphaltene Precipitation from Crude Oils: A Case of Neglected Kinetic Effects," *Energy & Fuels*, vol. 23, no. 7 , pp. 3681-3686, Jul. 2009.

[2] M.Hoepfner, C.V.B Favero and H. S. Fogler, " Understanding the Instability of Asphaltenes," Oral at Petrophase 2011, London.

(29h) Study of Heavy Crude Flows in Pipelines with Electromagnetic Heaters

Ricardo Dunia, Chemical Engineering Department, The University of Texas at Austin, Austin, TX and Thomas F. Edgar, Department of Chemical Engineering, The University of Texas at Austin, Austin, TX

The electromagnetic heating of heavy crude oil in cylindrical pipes represents a novel technique to reduce fluid viscosity and diminish the cost required for its transportation [1]. The efficiency of electromagnetic heaters (EH) depends of the material to be heated and the reflective electromagnetic properties of the enclosure. In that way losses can be minimized and the energy generated by the electromagnetic field source can be readily absorbed by the processing material. The distribution of electromagnetic heating devices in pipelines between oil producers and consumers may provide a considerable advantage when compared to the high operational costs of fired-heaters and heavy equipment pump stations [2]. Furthermore, electromagnetic heaters can be placed within pump stations to reduce the crude oil viscosity. The figure below illustrates the use of EH to increase the distance between oil pump stations. Notice how the pressure profile can be impacted by the placement of an electromagnetic heater to the point that fewer pumping stations may be required between producers and consumers of heavy oil.

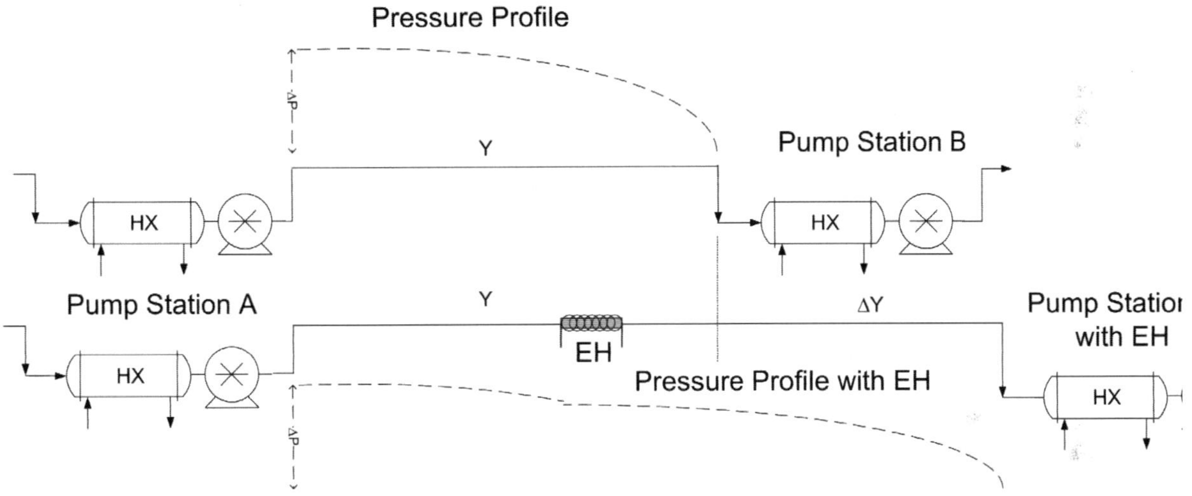

Figure: Crude oil transportation diagram and pipe pressure profiles. Top diagram shows the case of no intermediate heating between pump-stations. The energy loss due to high viscosity at low temperatures forces a separation of only Y miles between stations. The installation of an electromagnetic heater (EH) between pump-stations (bottom diagram) allows an extra separation of DY miles between stations, reducing the number of stations and pump head required for significant transportation distances.

In this study the oil viscous fluid momentum and energy balances, which include the effects of electromagnetic heating, variable viscosity and fluid dielectric properties, are solved in cylindrical coordinates [3]. The electromagnetic energy absorbed by oil is converted into sensible heat, which could significantly reduce the viscosity of heavy crude fluids. This drop in fluid viscosity diminishes flow pressure losses in pipelines, which potentially reduces the number and size of pumping stations between oil producers and consumers.

Different pipe materials are considered here to determine the effect of their dielectric properties in the fluid flow. Simulation results show that pipe materials with large electromagnetic absorption tend to attenuate the pipe interior electromagnetic field, which reduces the direct warming of the fluid. This significant reduction in the direct fluid heating suggests that pipes made by transparent electromagnetic materials are preferred for this type of applications.

The simulations presented in this work provide a first step in the estimation of electromagnetic heating for heavy crude transportation. An economic analysis of using electromagnetic heaters was made to maximize the distance between pump stations in a long pipeline stretch. The simulation results demonstrate that the use of electromagnetic heaters increases by 30% the distance between pump stations.

References

[1] Carrizales, M.; Lake, L. W. Two-Dimensional COMSOL Simulation of Heavy-Oil Recovery by Electromagnetic Heating. 2009.

[2] Martinez-Palou, R.; Mosqueira, M. d. L.; Zapata-Rendon, B.; Mar-Juarez, E.; Bernal- Huicochea, C.; Clavel-Lopez, J. d. l. C.; Aburto, J. Journal of Petroleum Science and Engineering 2011, 75, 274–282.

[3] Dunia, R.; Edgar, T. Use of Electromagnetic Heaters for Heavy Crude Oil Transportation. Submitted to Energy & Fuels 2012.

(41a) The Impacts of Biomass Particle Size During Air- and Oxy-Cofired Combustion

Melissa Holtmeyer[1], Gengda Li[2], Benjamin M. Kumfer[1], Shuiqing Li[2] and Richard L. Axelbaum[1], (1)Energy, Environmental & Chemical Engineering, Washington University in St. Louis, St. Louis, MO, (2)Department of Thermal Engineering, Tsinghua University, Beijing, China

The cofiring of biomass in pulverized coal boilers for large-scale power generation can only be a viable option if current combustion standards of stability, reliability, emission reduction, and fuel conversion efficiency are maintained and/or improved. While biomass fuels can be highly variable and differ significantly from coal, generally, they have increased volatile matter content and increased particle sizes, which are due to both fuel preparation methods as well as the structure of the biomass material. These two characteristics can significantly impact the structure of the volatile flame. The volatile flame zone is dominated by the combustion of volatiles in the near burner region. The length and location of the volatile flame envelope is also important not only to flame stability, but also to the formation of pollutants including NOx and SOx.

In this work, pulverized coal and biomass cofired flames are investigated under both air and oxyfuel combustion conditions using both experiment and Computational Fluid Dynamics (CFD) to determine the effects of volatile fraction, fuel composition, and particle size on the length of the volatile flame. Volatile flame length is inferred from experimentally obtained gaseous species concentrations for CO and CO_2 in a 35 kW combustion facility. The numerical study evaluates the impacts of these variables, both individually and combined, on volatile flame length.

This study shows the length of the volatile flame envelope is sensitive to the location of volatile matter release and the amount of volatiles available. For larger particles with high momentum and long heating times, they may pass through the oxygen-lean flame zone and release volatiles outside the volatile flame envelope into areas of higher oxygen. Particle breakthrough of the volatile flame can result in reduced fuel concentration in the near burner fuel-rich zone, which can reduce the volatile flame length. Increased volatile matter content, characteristic of biomass fuels, can lead to increased volatile flame length if all volatiles are released in the near burner region. However, if particles breakthrough the volatile flame envelope, then shorter than expected flame lengths may result. The volatile matter content of the particles that have broken through the flame envelope is determined using thermogravimetric analysis (TGA). Particles are sampled using a nitrogen-quenched particulate matter sampling probe downstream of the volatile flame.

(41b) Oxy-Fuel Combustion Optical Flame Analysis and Characterization

Casey S. Carney[1,2], Rigel Woodside[2], Danylo Oryshchyn[2], Stephen J. Gerdemann[2] and Thomas Ochs[2], (1)URS Corporation, Albany, OR, (2)National Energy Technology Laboratory, U.S. Department of Energy, Albany, OR

Oxy-Fuel Combustion Optical Flame Analysis and Characterization

C. S. Carney[1,2], R. Woodside[1], S. J. Gerdemann[1], D. Oryshchyn[1], T. Ochs[1]

[1]National Energy Technology Laboratory, 1450 Queen Ave SW, Albany, OR 97321

[2]URS Corporation, P.O. Box 1959, Albany, OR 97321-0521

Oxy-fuel combustion uses conventional technology for pulverized coal combustion, generating a flue gas composed primarily of carbon dioxide and water. While conventional air-fired coal combustion flame behavior is fairly well understood and optimized, the absence of nitrogen and consequent increase of carbon dioxide fraction and mass flow differences in oxy-fuel combustion results in differing flame behavior. Two notable examples are flame temperature and heat transfer within the boiler radiant zone. A unique suite of optical instrumentation devices has been designed for the purpose of oxy-fuel combustion flame study: two monochromators, four total radiometers, two fiber optic CCD spectrometers, and a video camera. These instruments view the flame via optical ports mounted along the flame axis – some along the burner centerline, and some 16" below the centerline. This system is capable of analyzing the following flame components: temperature, radiance, stability, morphology, and radical species. The radiometric components of the system will tend to observe emitted radiation from the soot particles, which will be used as a proxy for the flame's temperature characteristic.

The monochromator records irradiance vs. wavelength over the range of 800-5000 nm. Average radiant particle temperature can be calculated by fitting the Planck blackbody radiation curve to the shape of the spectrum over an optimized wavelength range for scan time minimization. Monochromators are set at the same axial distance from the burner, with one viewing the flame centerline and the other viewing 16" below centerline. The radiometers record total radiant flame intensity over the range of 200-7000 nm and are located along the burner centerline axis, allowing for a flame intensity profile along the flame length. The radiometers are also able to indicate flame oscillatory behavior. The CCD spectrometers provide instantaneous and continuous radiance vs. wavelength spectra in the range of 230-1100 nm. CCD spectrometer data are applied to calculate flame temperatures in their viewing area using the three-color method (evaluating the radiance ratio at three wavelengths). Flame radical species can be identified by these spectrometers, which are of use for combustion diagnostics. Rapid scan times enable the spectrometers to also indicate oscillations in the flame. Flame stability and morphology are monitored with a high speed video camera with a 90° field of view that has been customized for high temperature flame use. This camera is situated so that the entire length of the flame is recorded.

As a first test, this optical flame measurement system has been installed at the Jupiter Oxygen Burner Test Facility (15 MWth) in Hammond, IN. The test facility also contains various non-optical devices that further aid flame characterization such as thermocouples installed in inter-tube webbing and "chordal" thermocouples which can be used to determine boiler tube heat flux. Online flue gas compositional analysis records combustion-product volume-fraction composition at the boiler exit. The optical analysis system was operational for shakedown and burner tuning for both conventional air firing and oxy firing. Results are used both for flame characterization and comparison to existing computational fluid dynamics boiler simulations, including radiant zone temperature and heat transfer mapping.

(41c) Oxy-Fuel Combustion Characteristics of Victorian Brown Coal

Lian Zhang, Jian Zhang, Fiona Low and Wirhan Prationo, Chemical Engineering, Monash University, Melbourne, Australia

This paper introduced the R&D works in Monash University regarding the oxy-fuel combustion of Victorian brown coal. Victorian brown coal is the single largest source meeting >85% of the electricity in Victoria, Australia. Its high-moisture content (up to 65 wt%) however results in a notoriously high carbon footprint for the existing air-fired power generations burning this coal. A vareity of lab-scale facilities including drop-tube furnace and flat-flame burner reactor have been used to assess the combustion characeristics of both dry and wet brown coal in O2/CO2, as well as the ash formation mechanisms particularly upon the recirculation of impure flue gas. A disctinct flame propagation has been confirmed for brown coal in O2/CO2. In combination with the predominant CO2, the inherent moisture in coal played an important role in coal ignition, volatile flame intensity and char burn out. Pre-drying of brown coal is crutical in intensifying its combustion in a lower oxygen fraction. The organically bound metals are exclusively present in brown coal, the vaporisation extent of which upon combustion are promoted by coal pre-drying and use of high oxygen fraction. Apart from the knowledge achieved from lab-scale testing, the pilot-scale testing results for brown coal in a 3MWe furnace will also be introduced.

(41d) Heat Integration in Coal Based Oxy-Combustion Power Plants

Chao Fu, Department of Energy and Process Engineering, Norwegian University of Science and Technology, Trondheim, Norway and Truls Gundersen, Energy and Process Engineering, Norwegian University of Science and Technology, Trondheim, Norway

Oxy-combustion is a competitive technology to enable the capture of CO_2 from coal based power plants. The reduction in power efficiency and the increment of investment cost are less than (or at least comparable to) other CO_2 capture technologies. In addition, the possibility of co-capturing SOx and NOx is attractive when oxy-combustion is applied in coal based power plants. The core concept of oxy-combustion is to use high purity oxygen instead of air for the combustion process so that the flue gas is composed mainly of CO_2 and H_2O. The CO_2 can be separated by condensing the H_2O and then purified by chilling. After being compressed to a dense phase, the CO_2 can be transported to appropriate storage sites.

The oxy-combustion technology is ready for demonstration in both new and retrofitting power plants [1]. The energy penalty and investment cost related to CO_2 capture are the main barriers to implement this technology. Commercially available air separation technologies for high volume oxygen production which can be applied in oxy-combustion plants are based on cryogenic distillation. When a traditional double-column distillation cycle is applied for the supply of O_2 in a coal based oxy-combustion power plant, the total thermal efficiency penalty related to CO_2 capture is around 10.3% points (based on the higher heating value) [2]. The cryogenic air separation unit (ASU) contributes 6.6% points and the CO_2 compression and purification unit (CPU) contributes 3.7% points. The total efficiency penalty is expected to be reduced to less than 6% points in the future [3], while the theoretical minimum is 3.4% points when all the unit operations in the ASU and the CPU are assumed reversible [2].

For oxy-combustion power plants, the required O_2 purity is around 95 mole% and the supply pressure is slightly above atmospheric pressure. These new specifications on the O_2 product provide considerable opportunities for improving the ASUs. When the technology is shifted from thermally-coupled distillation to vapor recompression distillation, the energy consumption can be reduced by 7.5% [4]. In addition, when the principle of distributed reboiling is applied and combined with vapor recompression distillation, the energy consumption can be reduced by 17.1%. Thus the thermal efficiency penalty related to CO_2 capture can be reduced from 10.3 to 9.2% points.

Both the ASU and the CPU are sub-ambient separation processes. Mechanical work is consumed by the compressors in the refrigeration processes. The compression heat is generally removed by cooling water to lower the operating temperature, thus reduce the work consumption in the compressors. An alternative option for the compression heat from the ASU and the CPU is to be integrated with the steam cycle by partially preheating the boiler feedwater, so that less steam is extracted. A preliminary study shows that such integration increases the thermal efficiency by around 1.5% points [5], thus the efficiency penalty related to CO_2 capture is expected to be reduced to around 7.7% points.

This paper presents a detailed study on the integration of the compression heat with the steam cycle. The integration potential is first investigated in the case when interstage cooling is applied in the compression processes. In this case, the compression work in the ASU and the CPU is not influenced by the integration. Then fully and/or partially adiabatic compression is applied instead of interstage cooling, so that the temperature of the compression heat is lifted. The upgraded compression heat can be used to preheat the boiler feedwater in the steam cycle to a higher temperature, thus less steam will be extracted at the higher pressure levels. However, more work will be consumed in the compression processes due to higher operating temperatures. In addition, the compressor efficiency should be lower in this case since the mass flow through the compressors is smaller at higher temperatures. Thus the integration problem becomes much more complex when fully and/or partially adiabatic compression is applied. Such integration issues will be discussed in this paper.

References

[1] IEAGHG, Greenhouse News, 2011: NO. 104.

[2] C. Fu, T. Gundersen. Heat integration of an oxy-combustion process for coal-fired power plants with CO_2 capture by pinch analysis. *Chem. Eng. Transactions,* 2010: 21, 181-186.

[3] J.P. Tranier, R. Dubettier, N. Perrin. Air separation unit for oxy-coal combustion systems. 1st International Oxyfuel Combustion Conference, 7-11 September, 2009, Cottbus, Germany.

[4] C. Fu, T. Gundersen, D. Eimer. Air separation. GB patent, application number: GB1112988.9, 2011.

[5] T. H. Zeiner. Process integration potentials in coal based power plants using oxy-combustion. Master thesis, Department of Energy and Process Engineering, Norwegian University of Science and Technology, 2012.

Climate Change and Sustainable Development Legal Update

Monday, October 29, 2012

ENVIRONMENTAL & CARBON REPORTING & DISCLOSURE:

NOT BUSINESS AS USUAL

By

John. P. Fillo, Ph.D. CPEA

EXTENDED ABSTRACT

Climate change and carbon risk management are complex environmental and societal issues, as well as being fundamental to business sustainability. Increased regulation and reporting of environmental and carbon exposures, liabilities and risks are driven by national, regional, state/provincial and local requirements in the U.S. and abroad. These obligations have evolved from participation in voluntary initiatives to legislative and regulatory-driven mandatory actions. While reporting of greenhouse gas (GHG) emissions is only one component of these requirements, statutory financial disclosure requirements for publicly held companies and even country-specific requirements for environmental/ sustainability reporting extend these obligations to address strategy, operations, business risk and opportunity management, legal compliance and supply chain issues, as well as disclosures in financial statements and reports.

This presentation will provide an overview of and insight into the complexity of these requirements and their relevance to the chemical engineering profession. While it will provide a brief overview of the mandatory 'environmental' reporting programs and requirements as a baseline, the focus will be on statutory, industry and voluntary environmental, carbon and sustainability reporting and disclosure in the U.S. and internationally. These requirements and voluntary venues represent complex carbon reporting and disclosure requirements, and can extend beyond just simply reporting a GHG inventory. Such disclosures often require that corporate strategy, risk and opportunity management, and programs and initiatives be disclosed as a basis for assessing the veracity of such disclosures.

Mandatory Reporting & Disclosure

The U.S. is faced with a patchwork quilt of GHG reporting and disclosure requirements at the Federal, regional and state levels:

- The U.S. EPA's GHG Mandatory Reporting Rule, for which annual reporting of Scope 1 GHG emissions is required for numerous industry sectors.

- Regional GHG programs are a work in progress, with the most active being the Regional GHG Initiative (RGGI), the first market-based regulatory program in the U.S. to reduce GHG emissions.
- Several states have or are developing GHG reporting and regulatory programs, with California's Global Warming Solutions Act (AB-32) enacted, with GHG inventory reporting and verification requirements in place and cap and trade program requirements recently issued in 2012.

Additionally, U.S. Federal sustainability and carbon planning and reporting requirements have been in place since late 2009 under Executive Order 13514, *Federal Leadership in Environmental, Energy and Economic Performance*. The Executive Order requires annual reporting of Scope 1, 2 and 3 GHG emissions, as well as setting and achieving goals for energy use and GHG emissions reductions, and improvement of broad-based sustainability performance.

And our neighbors to the North have GHG reporting requirements, as well. Canada has Federal annual GHG emissions reporting requirements. Several provinces require industry-wide reporting of GHG emissions (e.g., Alberta, British Columbia, Quebec and Ontario), with some requiring formal verification of these emissions above stipulated thresholds.

Environmental, social and carbon reporting and disclosure requirements are typically embedded in securities reporting regimes throughout the world:

- Environmental and carbon risk reporting and disclosure are mandated by the U.S. Securities and Exchange Commission under Regulation S-K. Similar statutory requirements exist in Canada through Provincial securities commission requirements.
- European companies are required to include non-financial information in their annual and consolidated reports per the European Accounts Modernization Directive, and country-specific enabling legislation that can include requirements for reporting and disclosure of environmental, social and governance information.
- Country-specific requirements can mandate environmental and/or sustainability reporting (e.g., Japan, Brazil).

And while in the early stages of development, additional climate change and sustainability related reporting and disclosure mechanisms are being developed, which have the potential to become mandatory in the future:

- Global carbon disclosure and attestation standards, including those from the Climate Disclosure Standards Board, provide guidance on carbon risk disclosure for companies that are expected to follow International Financial Reporting Standards.

- The International Integrated Reporting Council is leading the development of a global framework for corporate reporting that demonstrates the linkages between an organization's strategy, governance and financial performance, and the social, environmental and economic context within which it operates.
- The Sustainability Accounting Standards Board, a US based, non-profit organization, intends to establish industry-based sustainability standards for the recognition and disclosure of 'material' environmental, social and governance impacts by companies traded on U.S. stock exchanges.

Voluntary/Industry Programs

While mandatory programs developed over the past few years in the U.S., GHG reporting and disclosure has been driven by a variety of voluntary programs and initiatives. These include, but are not limited to:

- The Carbon Disclosure Project, an independent not-for-profit organization, headquartered in the UK that seeks disclosure on climate change business risks and opportunities and GHG data from the world's largest companies since 2000. Reporting programs have extended into the areas of supply chain, water, and a cities project, as well.
- The Climate Registry, established in North America to measure, report and verify GHG emissions consistently across industry sectors.
- The Carbon Principles is a collaboration between investor-owned utilities, banks and NGOs to provide a consistent approach for banks and their U.S. power clients to evaluate and address carbon risks in financing electric power projects.
- The National Association of Insurance Commissioners has developed a reporting and disclosure framework that applies to insurers with aggregate annual premiums over $300 million, who are required to submit annual climate risk reports.

And there are additional voluntary venues, including GHG mitigation project development through carbon offset registries (e.g., Climate Action Reserve, Verified Carbon Standard, and others) and industry trade organization programs (e.g., Global Environmental Management Initiative, Asia-Pacific Partnership on Clean Development and Energy, and others).

Sustainability Indices

Sustainability indices can be an important venue through which to achieve reporting, disclosure and recognition. While these are frequently voluntary, representativeness, completeness, consistency, accuracy, and balance of disclosures speaks to corporate credibility and a company may be evaluated even if you choose not to formally participate. Corporate sustainability reporting is typically one of many criteria utilized to rank the included companies. Such indices include but are not limited to the Dow Jones Sustainability Index, FTSE4Good Index, Calvert Social Index, Morgan Stanley KLD 400 Social Index, the Newsweek Green Rankings and others.

Implications

And are these relevant to the chemical engineering profession? They most certainly are. For a chemical engineer whose career has evolved down a path of EHS/sustainability management or into the legal profession, they will be directly involved in the reporting and disclosure process. For the chemical engineer who is focused to product/process R&D or design and engineering, understanding these issues is essential to assure they are addressed progressively through the process and product development lifecycle. Engineering efficiencies and pollution prevention into processes and products, and assuring continuous improvement, is essential so as to improve a corporation's top and bottom line, and to ensure that the necessary information is available to support these required disclosures.

(77b) Epa's Clean Air Act Authority to Regulate Greenhouse Gases

Mary Ellen Ternes, McAfee& Taft, Oklahoma City, OK

Ever since the passage of the 1990 Clean Air Act Amendments, regulation of greenhouse gases has been controversial. Throughout the George W. Bush administration, EPA creatively avoided adopting Clean Air Act regulations governing greenhouse gases. Then, with the 2007 landmark United States Supreme Court case, Massachusetts v. EPA, the EPA was required to directly address greenhouse gas emission sources pursuant to the Clean Air Act. In 2009 and 2010, EPA initiated several rulemakings governing greenhouse gas emissions, including the Greenhouse Gas Reporting Rule, the Endangerment and Cause and Contribute Finding, the mobile source emission standards (the "Tailpipe" rule), the Tailoring and Timing Rules. Throughout this period, the U.S. House of Representatives entertained several proposals to eliminate EPA, combine EPA with the Department of Energy or simply modify the Clean Air Act to remove any authority to regulate greenhouse gas emissions. Then, in 2012, the D.C. Circuit reviewed all the legal challenges to the suite of greenhouse gas rulemakings and issued its decision. This presentation will review the D.C. Circuit's decision regarding the validity of EPA's greenhouse gas regulations, impacts from the decision, and implications for future policy.

(77c) Synthesis of CaO-Scaffold Adsorbent with Superior Stability for High-Temperature CO_2 Capture

Ming Zhao, School of Chemical & Biomolecular Engineering, The University of Sydney, Sydney, Australia, Nicholas H. Florin, Chemical Engineering, Imperial College, London, London, United Kingdom, Paul S Fennell, Chemical Engineering and Chemical Technology, Imperial College London, London, United Kingdom and Andrew T. Harris, Chemical Engineering, University of Sydney, Sydney, Australia

Introduction

Reducing greenhouse gas emissions to the atmosphere is a key environmental issue facing the world in general. Carbon dioxide (CO_2), primarily from the combustion of fossil fuels for energy production, is the most common greenhouse gas emitted by human activities and the one with the most significant impact on our climate. Calcium looping is one of the most promising technologies for both pre- and post-combustion CO_2 captures at a wide temperature range (400-900 °C) due to the high CO_2 carrying capacity and economical availability [1]. However, the natural CaO-based sorbents (limestone, calcite, dolomite, etc.) and even the synthetic species with tuned structures are suffering from a rapid decay in activity because of a serious sintering and fraction deduction, both of which happen to a pure CaO sorbent [2].

To solve the problem, we developed a sol-gel method to graft CaO onto an inert scaffold (silicon oxide) in order to mitigate sintering and enhance the mechanical properties. This method gives a new approach to a porous and stable CaO-based sorbent that would not only improve CO_2 capture but also enable the introduction of extra metallic sites to form a sorbent/catalyst composite.

Experimental and results

A saturated ethanol solution of tetraethoxysilane (TEOS) and $Ca(NO_3)_2$ was added dropwise into the mixture of 2.5 M ammonium hydroxide, hexadecyltrimethyl ammonium bromide (CTAB) and excess Na_2CO_3. The mixture was gelled and aged at 60 °C over a total of 24 h. The precipitate was washed, dried and calcined at 550 °C in air for 4 h. This CaO-scaffold sorbent was prepared with $n_{Si}:n_{Ca}$ = 1:1. A commercial sorbent that is stored as micrometer-sized $CaCO_3$ (Sigma-Aldrich) was used for comparison.

$$Si(OC_2H_5)_4\,(TEOS) + 2\,H_2O \xrightarrow{OH^{-1},\ \Delta} SiO_2 \downarrow +4\,C_2H_5OH \tag{1}$$

$$Na_2CO_3 + Ca(NO_3)_2 \longrightarrow CaCO_3 \downarrow +2NaNO_3 \tag{2}$$

CTAB was used as a template agent due to its positively charged molecules and formed growing micelles by combining negatively charged hydrated silica that derived from hydrolysis of TEOS [3]. At a higher temperature (60 °C) than conventional MCM-41 synthesis condition (25 °C), the aggregation and precipitation of SiO_2 (Eq. 1) was observed much accelerated so that a co-precipitation together with the formation of $CaCO_3$ (Eq. 2) was initiated. The resultant solid product was found with a one-fold framework instead of separately grown mixture. The morphology imaged by SEM is shown in Fig. 1 where numerous spherical nano granules (20-50 nm) can be observed with regularly arrangement simultaneously forming porosity in nano scale. Therefore, this CaO-scaffold sorbent gave a much higher surface area (111.1 m^2/g by BET model) than a pure-CaO-based sorbent (mostly $<10\ m^2/g$). The local ESD spectra indicate an even distribution of Si and Ca that also reflects the one-fold framework. Free $CaCO_3$ phases are confirmed by powder XRD technique, which predicts a reactivity of capturing CO_2 molecules.

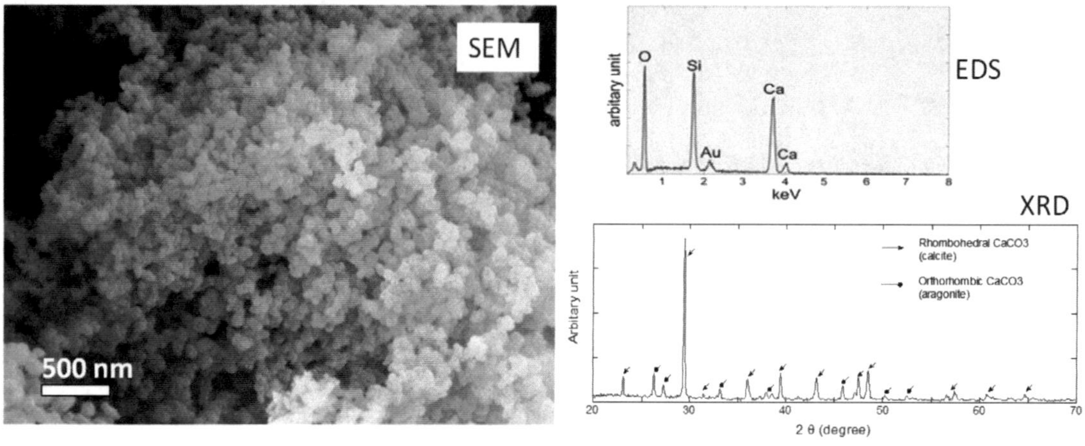

Fig. 1 Characterisation of the CaO-scaffold sorbent

The stability in multiple CO_2-capture-and-release runs was tested using a thermogravimetric analyser (TA Q5000IR). The sorbents were decarbonated in N_2 at 700 °C for 10 min followed by a cooling-down step to 600 °C. A 60-min carbonation in 60 vol.% CO_2 in N_2 then commenced. The decarbonation-carbonation process was repeated up to 13 cycles. The solid curves in Fig. 2 show the CO_2 carrying capacities (per unit sorbent) of the CaO-scaffold and commercial CaO sorbent, respectively, while the red dash line indicates the CO_2 uptake per unit CaO of CaO-scaffold, the value of which is different from the CO_2 uptake per unit sorbent due to the inert component. The new CaO-scaffold sorbent showed an excellent stability over multi-cycle tests despite the lower CO_2 uptake per unite sorbent compared to the commercial one. This stability reflected a superior ability to mitigate sintering deactivation that significantly decreases the practical potential of CaO-based CO_2 capture and is still investigated by current researchers. As shown by the red dash curve, the CO_2 binding capacity of free CaO on the new sorbent has surpassed the commercial one since the 2nd cycle and remained a nearly flat dropping track. This is likely attributed to the strong interaction between free Ca species with the scaffold [4].

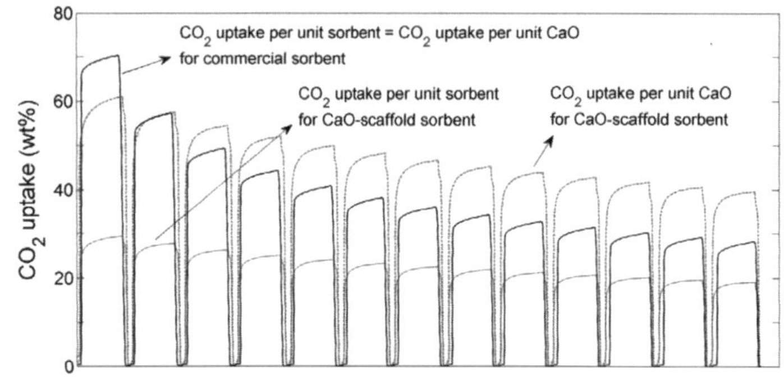

Fig. 2 CO_2-capture-and-release multiple tests using TGA

Conclusions

The demonstrated sol-gel method formed interaction between CaO and scaffold (SiO_2) and consequently a stable, porous, and ordered structure, which favored the sintering-resistant stability over multiple usages. Further investigations will be conducted for the deep understanding of the structure and optimization of the synthesis. The porous framework also provides space to house extra metallic catalyst, which would give a dual enhancement for hydrogen production from coal or biomass.

Acknowledgement. Authors are grateful to the Australian Research Council under DP0666488 and University of Sydney for International Program Development Fund.

References

[1]. Yang, Z.; Zhao, M.; Florin, N. H.; Harris, A. T., Synthesis and Characterization of CaO Nanopods for High Temperature CO_2 Capture. *Industrial & Engineering Chemistry Research* **2009,** *48*, (24), 10765-10770.

[2]. Florin, N. H.; Harris, A. T., Screening CaO-Based Sorbents for CO_2 Capture in Biomass Gasifiers. *Energy & Fuels* **2008,** *22*, (4), 2734-2742.

[3]. Cheng, C.-F.; Ho Park, D.; Klinowski, J., Optimal parameters for the synthesis of the mesoporous molecular sieve [Si]-MCM-41. *Journal of the Chemical Society, Faraday Transactions* **1997,** *93*, (1), 193-197.

[4]. Zhao, M.; Yang, X.; Church, T. L.; Harris, A. T., Novel CaO¨CSiO_2 Sorbent and Bifunctional Ni/Co¨CCaO/SiO_2 Complex for Selective H_2 Synthesis from Cellulose. *Environmental Science & Technology* **2012,** *46*, (5), 2976-2983.

In-Situ Synthesis of Useful Polyamines for CO_2 Capture from Piperazine

Omkar Namjoshi[1], *Yang Du*[1], *Han Li*[1,2], *Mark Goldman*[1], *Gary Rochelle*[1]

[1]The University of Texas at Austin, Austin, Texas, USA; [2]Tsinghua University, Beijing, PRC

Abstract

Concentrated piperazine (PZ) is considered to be a promising solvent for use in post-combustion CO_2 capture from coal-derived flue gas due to its fast rate of CO_2 absorption, high CO_2 capacity, and low degradation rates. However, this solvent has a limited solubility window at conditions corresponding to the operating temperature of a CO2 capture plant, limiting its use in commercial CO_2 capture systems. Blending PZ with a triamine analogue of PZ, such as aminoethylpiperazine (AEP), can widen the solubility window while maintaining many of the properties that make concentrated PZ attractive. With the exception of AEP, many of the triamine PZ analogues are expensive and are not commercially available. They can, however, be made in-situ in the regeneration section of the CO_2 capture plant by the selective reaction of PZ with an amine that forms an oxazolidinone (either a primary or secondary alkanolamine) as its initial thermal degradation byproduct.

The data suggests that the structure of the oxazolidinone is a strong function of reactivity with PZ. Increasing steric hindrance – such as increasing the chain length between the amino group or hydroxyl group or by adding alkyl functional groups to the carbon chain between the hydroxyl and amino groups – can reduce reaction rates by a factor of three to five. Secondary amines that form oxazolidinones react four times as fast as primary oxazolidinone-forming amines. Piperazine and solvent alkalinity losses are reduced significantly (~0.2%/week) once the oxazolidinone-forming amine is completely consumed, indicating that the synthesized amine blend is thermally stable.

Keywords: Carbon Capture, Amine Synthesis, Piperazine, Oxazolidinones

1. Introduction and Theory

A novel method for synthesizing triamines and tetramines in the regeneration section (stripping column) of an amine CO_2 capture plant was investigated. This process uses piperazine (PZ) or a PZ derivative and an amine whose initial degradation product is an oxazolidinone.

Oxazolidinones are the initial thermal degradation byproducts of primary or secondary alkanolamine with two or three carbon atoms between the oxygen and nitrogen atoms. They form when the amine carbamate (which, along with a protonated amine, is formed when CO_2 reacts with a primary or secondary amine) first undergoes electron abstraction followed by ring closing (Davis, 2009; Gouedard 2012). This scheme is shown in figure 1.1.

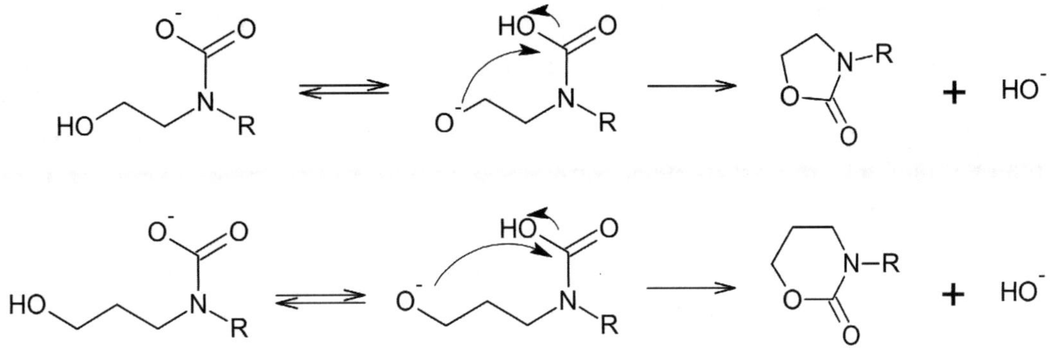

Figure 1.1. General scheme of oxazolidinone formation.

Primary and secondary amines can react rapidly with oxazolidinones to form substituted amine synthesis products. This general scheme is shown in figure 1.2.

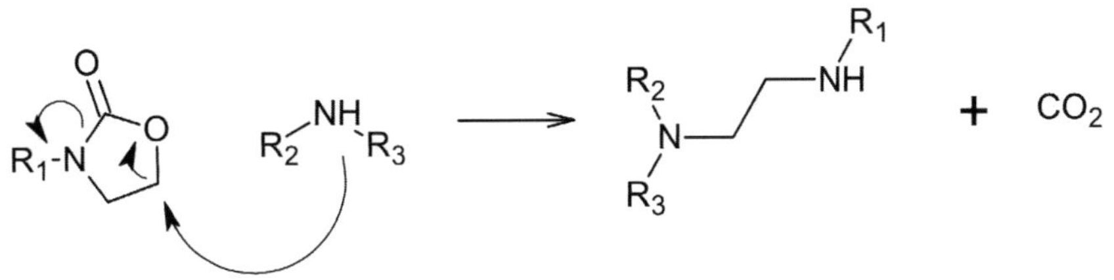

Figure 1.2. General scheme of reaction a primary or secondary amine with an oxazolidinone to form a synthesis product and CO2.

The oxazolidinone-forming amines studied in this paper include monoethanolamine (MEA), propanolamine (MPA), methylisopropanolamine (MIPA), 2-amino-2-methyl-1-propanol (AMP), and methylaminoethanol (MAE). Structures of these amines are shown in table 1.1. An example reaction scheme between PZ and MAE to form the synthesis product methylaminoethylpiperazine is shown in figure 1.3. The structures of the synthesis products formed by reaction with PZ are shown in table 1.2.

Table 1.1. List of the oxazolidinone forming amines used in this study

Amine Name	Abbreviation	Amine Structure	Structure of the Oxazolidinone
Monoethanolamine	MEA	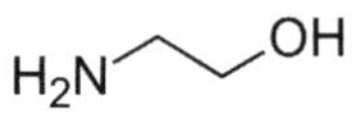	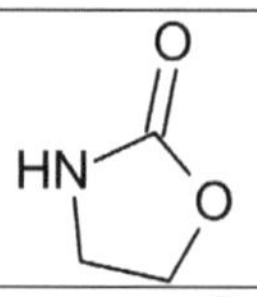
Methylaminoethanol	MAE	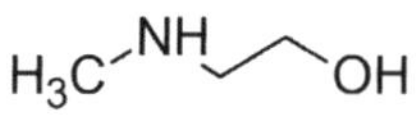	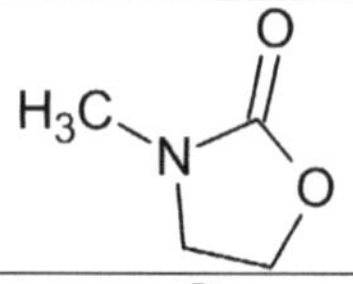
Monoisopropanolamine	MIPA	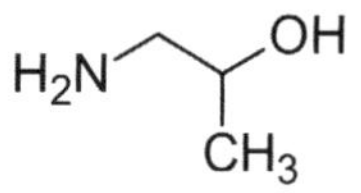	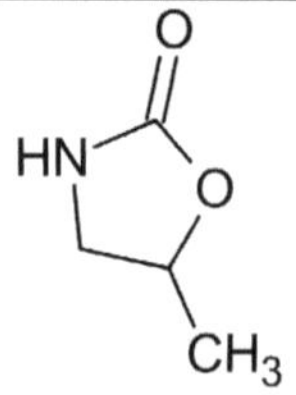
2-Amino-2-methyl-1-propanol	AMP	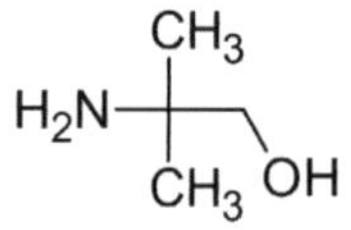	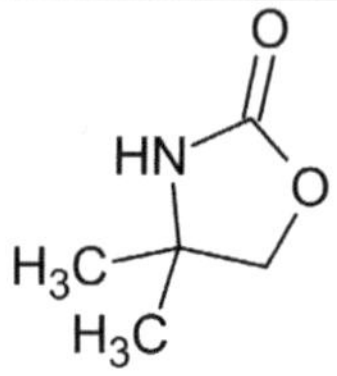
Monopropanolamine	MPA		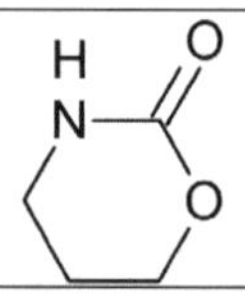

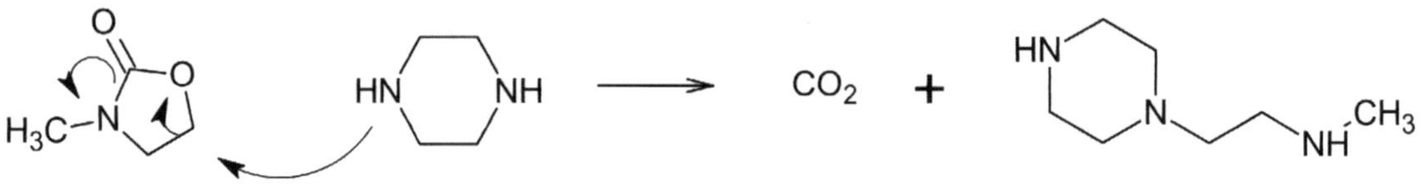

Figure 1.3. Formation of methylaminoethylpiperazine by reaction of PZ with MAE.

Table 1.2. Structures of the principal synthesis products seen in reaction between PZ and the oxazolidinone-forming amine

Amine	Triamine Synthesis Product	Tetramine Synthesis Product
Monoethanolamine	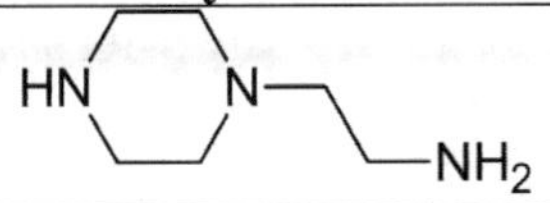	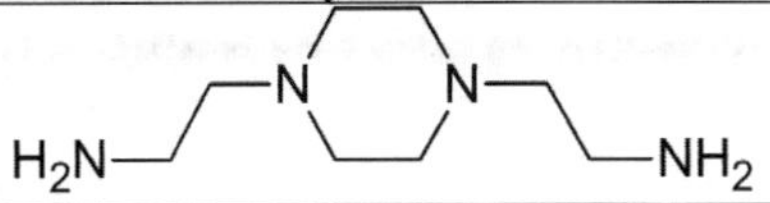
Methylaminoethanol	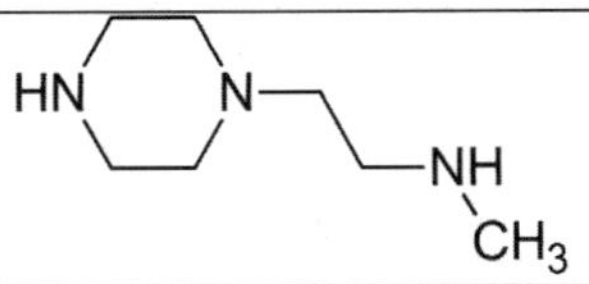	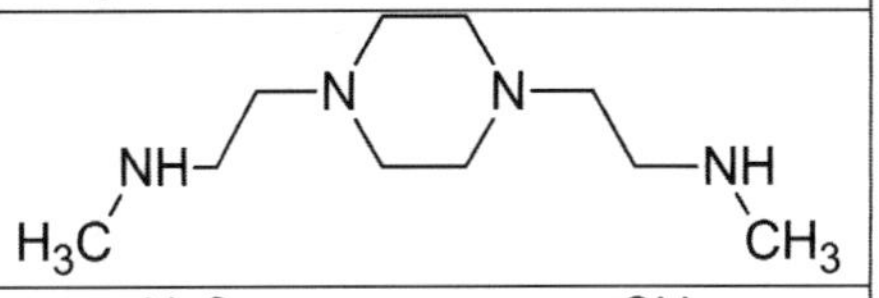
Monoisopropanolamine	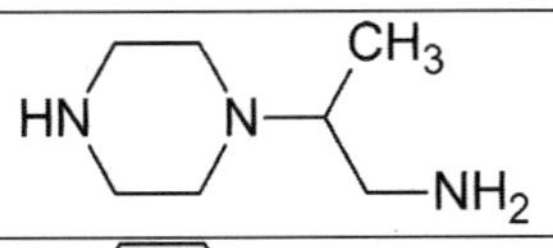	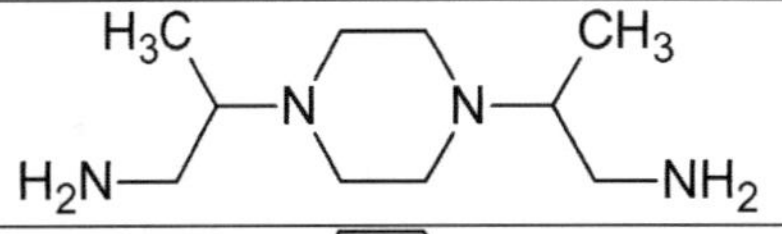
2-Amino-2-methyl-1-propanol	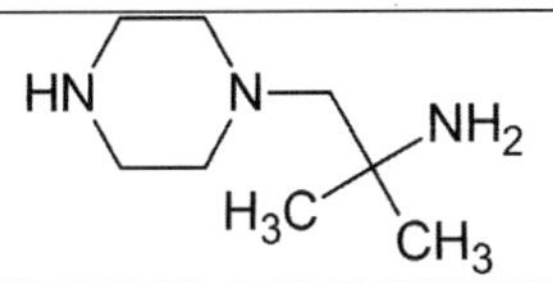	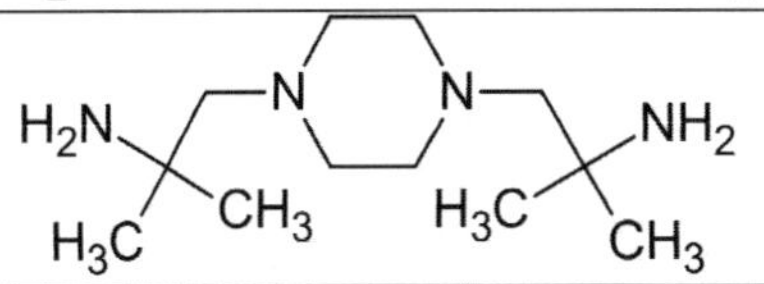
Monopropanolamine	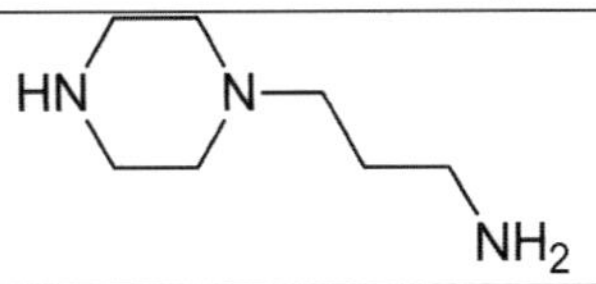	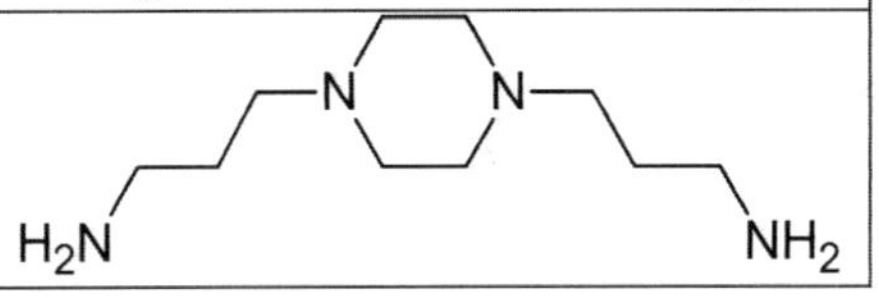

AEP has been studied as a potential capture solvent by Chen (Chen, 2011) and Du (Rochelle, 2012a; Rochelle, 2012b). 6 m AEP and 5 m PZ/2.3 m AEP both have lower cyclic CO_2 capacities than 8 m PZ but have competitive CO_2 absorption rates, slightly higher heats of absorption and, in the case of 5 m PZ/2.3 m AEP, have comparable thermal degradation characteristics as 8 m PZ. AEP, when blended with PZ, has better solid solubility characteristics than 8 m PZ, a key limitation of the concentrated PZ solvent. AEP (and other triamines and tetramines) are less volatile than PZ and thus will have reduced amine emissions from the CO_2 absorption column and/or the water wash column. Although AEP is commercially available, the other synthesis products, which are shown in in table 1.2, are not readily available. These synthesis products, however, can be produced in the regeneration section of the CO_2 capture plant.

2. Experimental Methods

Samples were prepared gravimetrically. CO_2 was sparged into the solution and measured gravimetrically. An initial CO_2 loading of 0.3 mol CO_2/mol alkalinity was used for each solution that was prepared. 7 ml of solution was loaded into 10 ml stainless-steel Swagelok® cylinders, which were then sealed and placed in convection ovens at set temperatures of 135 and 150 degrees C. These temperatures correspond to those encountered in the regeneration sections of proposed CO_2 capture plants using amine solvents. Cylinders were removed periodically throughout the experimental run.

A Dionex ICS-2100 chromatography unit was used to analyze for parent amine concentrations and synthesis product concentrations. Samples were diluted 10000X by mass. A Dionex CG17 50 mm guard column and a CS17 250 mm analytical column connected in series were used to carry out the separation 18.2 umho deionized water spiked with a gradient of methylsulfonicacid was used as the mobile phase. A Metrohm potentiometric autotitrator was used to estimate the loss of alkalinity in each sample. Samples were diluted 300X by mass and 0.2N H_2SO_4 was added until the pH reading reached approximately 3.9, corresponding to total protonation of all amine compounds present in the sample. The experimental methods are essentially similar to those of Freeman (Freeman, 2011).

A list of chemical suppliers and product purity is presented in table 2.1.

Table 2.1. List of chemicals suppliers and product purity

Chemical name	Purity	Supplier
Piperazine (PZ)	99	Sigma-Aldrich
Monoethanolamine (MEA)	99	Acros Organics
Monopropanolamine (MPA)	99	Acros Organics
Monoisopropanolamine (MIPA)	93	Sigma-Aldrich
2-Aminomethylpropanol (AMP)	99	Acros Organics
Methylaminoethanol (MAE)	99	Acros Organics

The supplied MIPA likely contains an isomer as its principal impurity (Bedell, personal communication, May 2012). This isomer is shown in figure 2.1 and can also participate in the amine synthesis like MIPA.

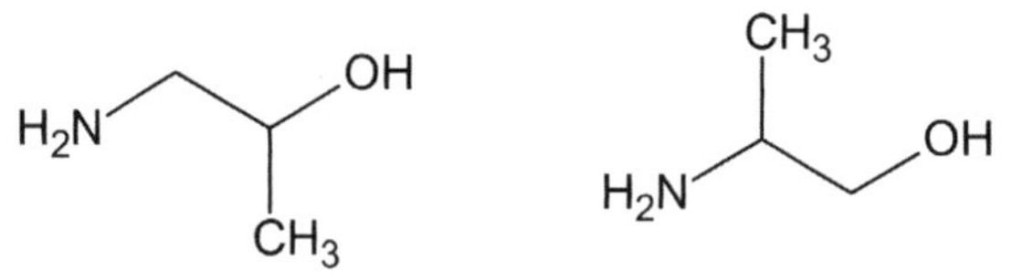

Figure 2.1. Isomers of MIPA; the principal product is on the left

3. Results and Analysis

Unified Calibration Curve

Standards were unavailable for several of the synthesis compounds made. The calibration data, however, for AEP and PZ can be modeled well using a single calibration curve. These calibration data were used to estimate the quantities of the synthesis products present in each sample; the curve is presented in figure 3.1.

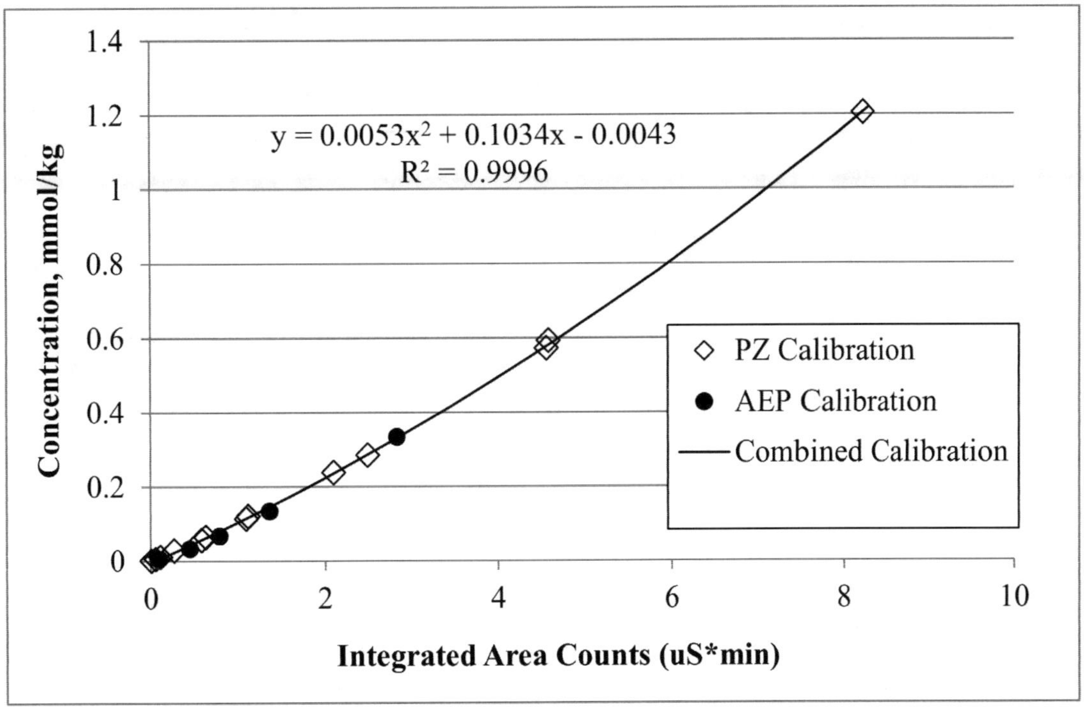

Figure 3.1. Unified Calibration Curve for PZ and AEP

Representative Chromatograms

A sample chromatogram of the synthesis reaction between 7 m PZ/2 m AMP initially at 0.3 mol CO_2/mol alkalinity after 1030 hours and 150°C is shown in figure 3.2. The parent amines and the synthesis product have been highlighted in the chromatogram.

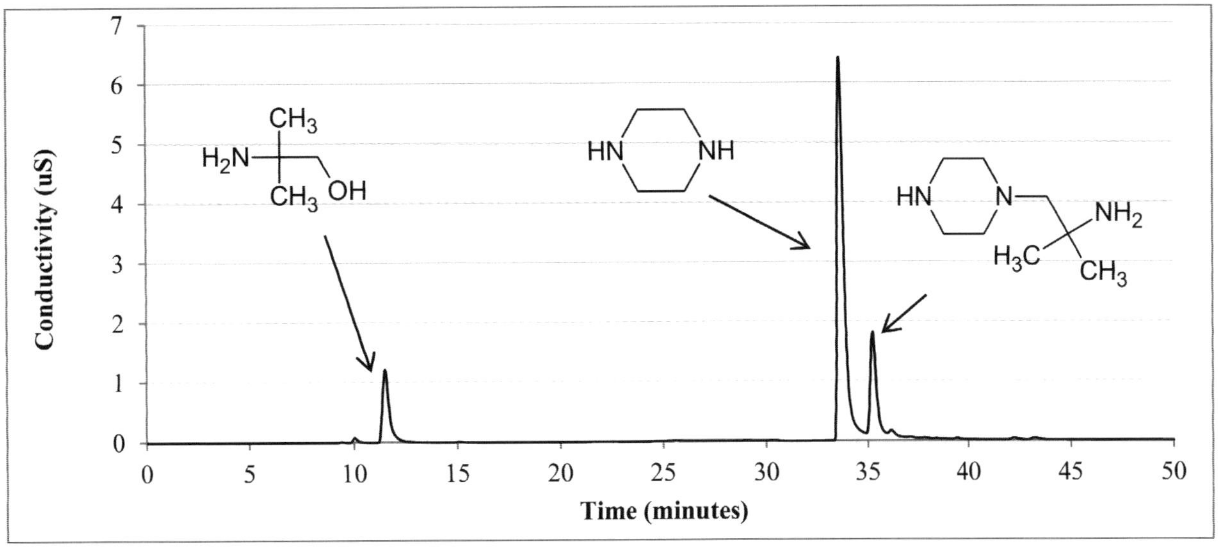

Figure 3.2. Sample chromatogram of AMP-PZ synthesis

A first order rate model was used to estimate the loss of the oxazolidinone-forming amine throughout the reaction run. Figures 3.3 and 3.4 suggest that this model approximates amine loss reasonably well.

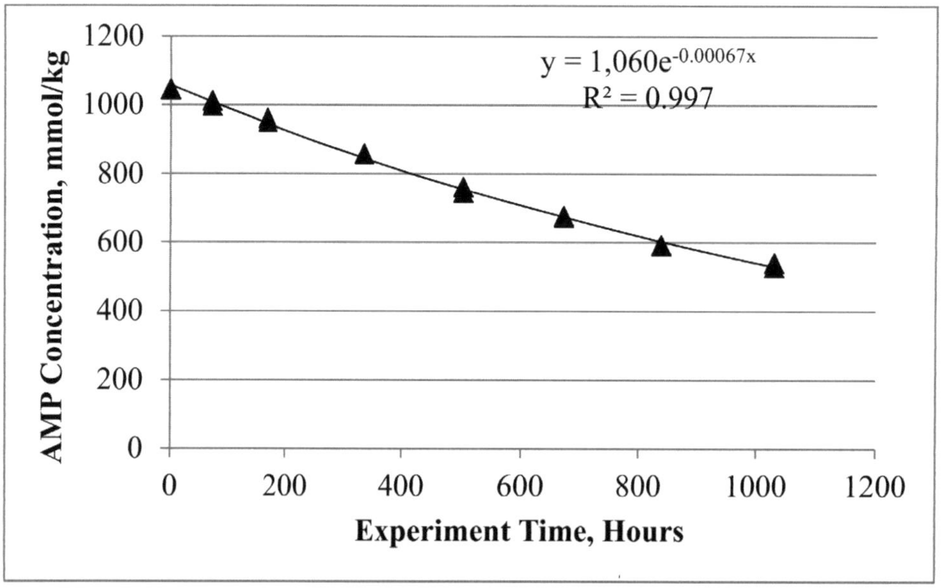

Figure 3.3. Concentration vs time plot of AMP in 7 m PZ/2 m AMP at 150°C, initially at 0.3 mol CO_2/mol alkalinity loading

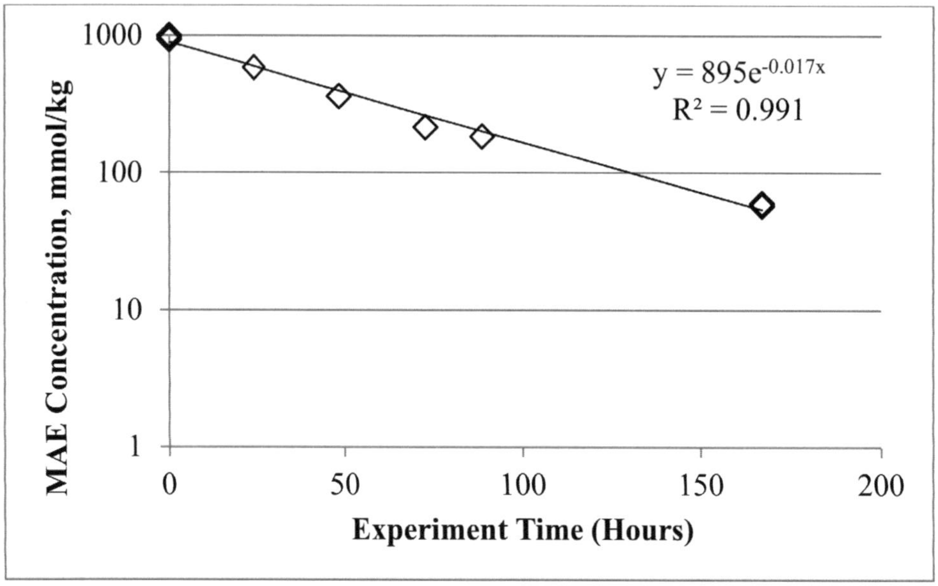

Figure 3.4. Concentration vs time plot of MAE in 7 m PZ/2 m MAE at 150°C, initially at 0.3 mol CO_2/mol alkalinity loading

The activation energy was estimated by using the Arrhenius relationship. Although this work only incorporates data from two different temperatures, data from Freeman suggests that thermal degradation of amines (among them oxazolidinone-formers) follow this behavior (Freeman, 2011).

Selectivity to the synthesis products is defined as the following:

$$\text{Selectivity}_t = \frac{\sum C_{\text{Synthesis Product},t}}{C_{\text{Oxazoidinone Former},i} - C_{\text{Oxazolidinone Former},t}} \tag{3.1}$$

Synthesis products can constitute both the triamine synthesis product and the tetramine dimer synthesis product shown in table 1.1.

Results Summary

Table 3.1 shows the summary of the results for the various amines tested.

Table 3.1 – Results Summary
Note: Degradation data apply to 150°C

Oxazolidinone-Forming Amine	**MEA**	**MPA**	**MIPA**	**AMP**	**MAE**
Amine Conversion (at 500 hours)	85%	45%	44%	28%	100% *(occurred between 167 and 340 hours)*
Synthesis Product Selectivity (at 500 hours)	69%	>90%	87%^	93%	>90%
Alkalinity Loss, %/week	2.0	1.7	1.5	1.0	5.0 (at 167 hours)*
First-order Amine Conversion Rate, 1/hr	0.0042	0.0012	0.00088	0.00067	0.017
Activation Energy, kJ/mol	160	150	130	140	100

* alkalinity of the solution was essentially unchanged after 167 hours (decreased by 0.2 %/wk).
^ also includes synthesis products from the secondary isomer

Analysis

Amine structure appears to have a strong effect on the rate of reaction with PZ. Increasing the steric hindrance of the oxazolidinone forming amine – either by lengthening the carbon chain from two to three carbons or by adding functional groups to the carbons in the alpha position relative to the amino or hydroxyl groups – reduces the rate of reaction with PZ by a factor of three to six. On the other hand, it appears as if oxazolidinone-forming secondary amines have a rate of reaction with PZ that is about four times as fast than the primary amines.

Alkalinity loss can be correlated with the oxazolidinone-forming amine conversion rate. Increased rates of alkalinity loss are indicative of the formation of ureas. Ureas can be formed from the synthesis product that is produced when an oxazolidinone-forming amine reacts with its own oxazolidinone. This is commonly encountered in MEA thermal degradation when the MEA-oxazolidinone synthesis product, hydroxyethylethylenediamine (HEEDA), can rapidly form a urea.

A key finding is that it that the alkalinity loss rate is reduced significantly after consumption of the oxazolidinone-forming amine – especially in the case of the PZ-MAE synthesis reaction. Once the MAE was consumed, the PZ loss was reduced to approximately 30 mmol/week and alkalinity loss was reduced by a factor of 25. This implies that the amine synthesis products and piperazine form a thermally stable mixture and therefore can be used at elevated temperatures in the regeneration section without significant amine loss. A similar phenomenon was noted in PZ-MEA synthesis, with PZ loss averaging about 10 mmol/week once the MEA was consumed.

4. Conclusions

a. First-order kinetics can accurately model the consumption (loss) of the oxazolidinone-forming amine when reacted with piperazine to form substituted triamine and tetramine piperazine analogues.
b. At 150°C, an initial amine concentration of 7 m piperazine and 2 m oxazolidinone forming amine, and at an initial loading of 0.3 mol CO_2/mol alkalinity, the oxazolidinone-forming amine consumption rate is the following:
   - Monoethanolamine: 0.0042 1/hr
   - Monopropanolamine: 0.0012 1/hr
   - Monoisopropanolamine: 0.00087 1/hr
   - 2-amino-2-methylpropanol: 0.00067 1/hr
   - Methylaminoethanol: 0.017 1/hr
c. The structure of the oxazolidinone-forming amine plays a significant role in its loss rate:
   - Increasing steric hindrance – such as increasing the chain length between the amino group or hydroxyl group or by adding alkyl functional groups to the carbon chain between the hydroxyl and amino groups – can reduce reaction rates by a factor of three to five.
   - Secondary amines that form oxazolidinones react four times as fast as primary oxazolidinone-forming amines.
d. Piperazine and solvent alkalinity losses are significantly reduced (~0.2 wt%/wk) once the oxazolidinone-forming amine is completely consumed, indicating thermal stability of the synthesized amine blend.

5. Acknowledgements

The authors wish to acknowledge the Luminant Carbon Management Program for funding.

6. References

Chen X. *Carbon Dioxide Thermodynamics, Kinetics, and Mass Transfer in Piperazine Derivatives and Other Amines.* The University of Texas at Austin. Ph.D. Dissertation. 2011.

Davis JD. *Thermal Degradation of Aqueous Amines Used for Carbon Dioxide Capture.* The University of Texas at Austin. Ph.D. Dissertation. 2009.

Freeman SA. *Thermal Degradation and Oxidation of Aqueous Piperazine for Carbon Dioxide Capture.* The University of Texas at Austin. Ph.D. Dissertation. 2011.

Gouedard C, Picq D, Launay F, Carrette PL. "Amine Degradation in CO2 Capture. I. A Review." *International Journal of Greenhouse Gas Control.* 2012;10:244-270.

Rochelle GT et al. "CO_2 Capture by Aqueous Absorption, Fourth Quarterly Progress Report 2011." Luminant Carbon Management Program. The University of Texas at Austin. 2012.

Rochelle GT et al. "CO_2 Capture by Aqueous Absorption, First Quarterly Progress Report 2012." Luminant Carbon Management Program. The University of Texas at Austin. 2012.

(132a) Characterization of the Oil Shale Core Pore Structure Before and After Pyrolysis

Pankaj Tiwari[1], Milind D. Deo[1], C.L. Lin[2] and J.D. Miller[2], (1)Chemical Engineering, University of Utah, Salt Lake City, UT, (2)Department of Metallurgical Engineering, University of Utah, Salt Lake City, UT

The pyrolysis of oil shale to produce transportation fuels is a complex process. The organic matter in the oil shale is tightly bound with varying mineral matrix. Several physical changes occur during the thermal conversion of kerogen in oil shale to produce hydrocarbon product. The creation of pore space during the pyrolysis is an important physical process. The study of the pore formation during the process can help in understanding the flow behavior of pyrolysis product within the internal network. In this paper, we report the effect of temperature on oil shale pyrolysis and product distribution, and study of the creation of pore volume during the process. Oil shale cores of 1" diameter from different depths of a single drill hole in the Unita Basin were used. Increase in the temperature resulted in higher weight loss and correspondingly the oil yield. However, the product distribution depends on the maturity of the material. Multiscale 3D X-ray CT imaging was performed to characterize and to analyze the nature of the pore network structure before and after pyrolysis. The CT scans of these cores at 42 micron voxel resolution displayed the distinguishable features of reaction products and source rock. Pyrolysis of organic rich core at high temperature produced large pore space during the process.

(132b) Application of an Integrated Multiphysics Simulator to Underground Coal Gasification

Charles Reid[1], John Nitao[2], Joshua White[2], Gregory Burton[2], Jeff Wagoner[2], David Camp[2] and Julio Friedmann[2], (1)Engineering, Lawrence Livermore National Laboratory, Livermore, CA, (2)Lawrence Livermore National Laboratory, Livermore, CA

Underground coal gasfication (UCG) is a problem rich with complexity and interconnected physical phenomena. It is also an opaque process, with limited operational data available and with the only clear picture of cavity evolution coming after the fact. A comprehensive UCG simulator has been developed, which is capable of accounting for the multiphysics complexity of UCG, to provide a better understanding of UCG dynamics (e.g. design, operation, monitoring, and hazards analysis). The simulator breaks UCG into six parts, each corresponding to a particular aspect of UCG physics, and each spawning a particular computational tool designed for a particular application. Each computational tool is tied to the others through a simulation manager, which manages grids, discrete elements, communication overhead, input/output, and transfer of information among components. The overarching UCG model that results is entirely novel, combining detailed models for thermal-hydrology, cavity gas, rubble inside the cavity, boundary evolution, wall reactions, and geomechanics into a single simulator and a single algorithm in an unprecedented way. The simulator is able to predict many quantities, including the evolution of the cavity boundaries, the potential for surface subsidence, and the impact of gas mixing and reaction in the cavity on cavity evolution. Several improvements have been made to the simulator over the past year, including rubble model improvements, identified in prior studies as a source of inaccuracy in comparisons with data for a simulation of Hoe Creek 3, improvements to the geomechanical collapse model, and fully coupled 3D cavity gas and rubble zone models. Results from simulations of the Rocky Mountain field tests are presented and compared with experimental data. Additional challenges and areas of interest for future development are identified.

(132c) On the Relationship Between Mechanical Specific Energy and Rate of Penetration

Yaneng Zhou, Department of Civil and Environmental Engineering, University of Pittsburgh, Pittsburgh, PA, **Jeen-Shang Lin**, National Energy Technology Laboratory, U.S. Department of Energy, And Department of Civil and Environmental Engineering, University of Pittsburgh, Pittsburgh, PA and Isaac K. Gamwo, National Energy Technology Laboratory, US Department of Energy, Pittsburgh, PA

A technique to measure the efficiency of a drilling operation consists of the examination of how the Mechanical Specific Energy (MSE) input is translated into the drilling rate of operation (ROP). This study investigates such relationship from laboratory-scale single cutter test to full drilling test. We initially carried finite element method modeling to provide insight on the mechanics involved. We then computed the mechanical specific energy using two volumes: the actual volume being cut and the volume computed based on the region swept by a cutter. Our finite element method analysis of a circular cutting by a sharp single disk PDC cutter provided a nearly constant mechanical specific energy when the rock failed in ductile mode. We have derived a general correlation of the mechanical specific energy versus the drilling rate of penetration. The correlation has been tested using data from the National Energy Technology single cutter ultra-deep drilling simulator at elevevated pressure of 138 MPa as well as data from rock cutting at ambient pressure. These comparative results will be presented.

Reactive transport modeling to evaluate interactions between pozzolan-amended wellbore cement and acid gas (CO_2 + H_2S) under typical acid gas sequestration conditions

Liwei Zhang, David A. Dzombak, and David V. Nakles
Department of Civil and Environmental Engineering, Carnegie Mellon University, Pittsburgh, PA, 15213-3890

Steven B. Hawthorne
Energy & Environmental Research Center, University of North Dakota, Grand Forks, ND, 58202

Barbara Kutchko, Christina Lopano, and Brian Strazisar
National Energy Technology Laboratory, Pittsburgh, PA, 15236-0940

Leopold Brunet, Li Li
Department of Energy and Mineral Engineering, Pennsylvania State University, University Park, PA, 16802

Abstract

Capture and subsurface co-sequestration of acid gas (H_2S and CO_2) is one sequestration approach that can reduce the emissions of both CO_2 and H_2S from combustion of fossil fuel. Before the implementation of acid gas co-sequestration, the potential of acid gas leakage along existing and abandoned wellbores at target sequestration regions needs to be evaluated. Reactive transport modeling is a powerful tool to mimic the interactions between wellbore cement and CO_2 + H_2S, so as to evaluate the potential of acid gas leakage along existing and abandoned wellbores. In this study, the interactions between pozzolan-amended wellbore cement and acid gas were simulated using the reactive transport modeling program CrunchFlow. Key simulation outputs include, for example, calcite weight percentage, pH, porosity, pyrite weight percentage and ettringite weight percentage from the exterior to the interior of the pozzolan-amended wellbore cement. To validate model simulation outputs, SEM-BSE and SEM-EDS analyses on real pozzolan-amended cement samples exposed to CO_2 and H_2S were conducted. Simulation results are consistent with SEM-BSE and SEM-EDS analyses results from real samples. Both model simulation and experimental results show that under the experimental exposure conditions ($T = 50^{\circ}C$, $P = 151$ bar, 21 mole % H_2S in 79 mole % CO_2 and samples immersed in 1 wt% NaCl solution), the alteration of pozzolan-amended wellbore cement porosity is not significant after 28 days of exposure.

1. Introduction

The integrity of wellbore cements in the simultaneous geologic sequestration of CO_2 and H_2S is an important issue in evaluating risk of leakage in co-sequestration projects (Jacquemet et al, 2008; Hawthorne et al, 2011; Kutchko et al, 2011; Jacquemet et al, 2012). Leakage of CO_2 and H_2S from abandoned wells at or close to the storage site is a potentially important leakage pathway for CO_2 and H_2S to escape from the storage site (Gasda et al, 2004; Theresa et al, 2009). CO_2 and H_2S both have potential to cause degradation of wellbore cement integrity, which could create CO_2 and H_2S leakage pathways along the well (Gasda et al, 2004).

The goal of this study is to develop a reactive transport model for the chemical alteration of wellbore cement by CO_2 and H_2S, so as to predict the changes in porosity and permeability of wellbore cement following exposure to a mixture of CO_2 and H_2S. The porosity and permeability changes of wellbore cement are important criteria for the evaluation of CO_2 and H_2S leakage risk along wellbores at the sequestration site.

2. Development of the reactive transport model using CrunchFlow

A multi-component reactive transport model to describe the interaction of CO_2 and H_2S with wellbore cement, and the resulting changes in well cement properties, is being constructed with use of the reactive transport model CrunchFlow (Steefel, 2009). The generalized process of interaction between CO_2 and wellbore cement is that CO_2 reacts with Ca-bearing phases (e.g., portlandite, $(CaO)_3SiO_2$, $(CaO)_2SiO_2$ and C-S-H) to produce $CaCO_3$. The generalized process of interaction between H_2S and wellbore cement is that H_2S reacts with ferrite phases (i.e., $(CaO)_4(Al_2O_3)(Fe_2O_3)$) and hydration products of ferrite phases to produce ettringite, gypsum and pyrite. The reactive transport model includes all CO_2-cement and H_2S-cement reactions involved in the cement-fluid interaction.

The mass conservation equation is the fundamental equation used in reactive transport modeling with CrunchFlow. The expression of mass conservation equation is (modified from Steefel and Lasaga, 1994):

$$\frac{\partial(\phi C_i)}{\partial t} = \underbrace{\nabla(\phi D_i \nabla C_i)}_{\text{Diffusion term}} - \underbrace{\nabla(\phi u C_i)}_{\text{Advection term}} \pm \underbrace{\sum_{r=1}^{N} v_{ir} R_{ir}}_{\text{Reaction term}} \qquad \textbf{(Eq. 1)}$$

where Φ is the porosity of the system, C_i is the concentration of species i, D_i is effective diffusion coefficient of species i, u is flow velocity, R_{ir} is the rth reaction rate that can produce or consume species i (the production or consumption of species i usually results in the dissolution or precipitation of a reactive mineral phase i), and v_{ir} is stoichiometric number of species i in the rth reaction.

Given the values or expressions of four parameters (Φ, D, u and R) in Equation 1, CrunchFlow can solve the mass conservation equation and generate concentration profiles of all species i as a function of time. It is important to note that the current modeling effort is for a static, or no-flow condition, and the advection term is set equal to 0 ($u = 0$).

3. Description of the cement-fluid system modeled

The system modeled is composed of two sub-systems: 1 wt % NaCl brine and the wellbore cement (Figure 1). The brine in the model is considered to be saturated with CO_2 and H_2S. The CO_2 concentration in brine is 0.904 mol/L and the H_2S concentration in brine is 1.356 mol/L, which are calculated based on Duan and Sun (2003) and Duan et al. (2007) under an exposure condition of 151 bar, 50°C and 21 mol % H_2S in H_2S-CO_2 gas mixture. The pH of the brine is calculated to be 2.9. It is assumed that the rates of CO_2 and H_2S consumption through reaction with cement are much less than the rates of CO_2 and H_2S supply to the brine from the supercritical H_2S-CO_2 mixture contacting the brine. Thus the CO_2 and H_2S concentrations in the brine phase are maintained at a stable level.

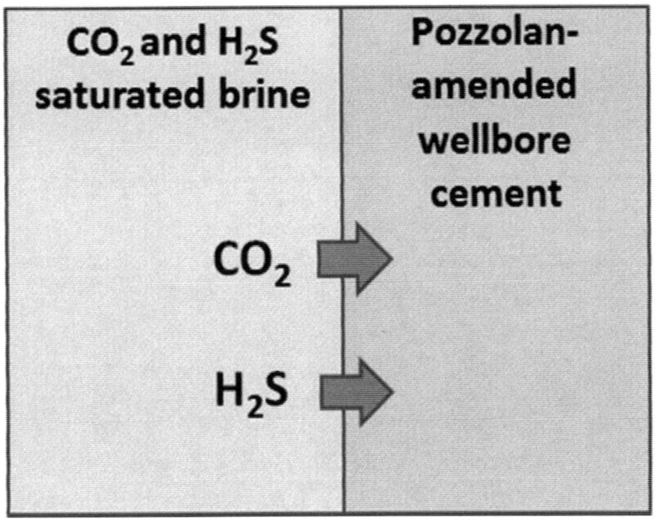

Figure 1: Conceptual depiction of the system modeled. CO_2 and H_2S saturated brine is next to the pozzolan-amended cement, with CO_2 and H_2S penetrating from the brine to the cement.

The cement in the model is considered to be pozzolan-amended Portland cement with 35 vol % pozzolan addition (nomenclature: 35 : 65 pozzolan-amended cement). The porosity of the 35 : 65 pozzolan-amended cement is 0.23, which was obtained from a BET test.

4. Results and discussion

The model was applied to some experimental data developed in related work. The experimental data include SEM-BSE imaging data (used to determine the position of the calcite-rich layer after exposure), SEM-EDS data (used to determine the position of the sulfur-rich layer after exposure), and BET test data (used to determine the porosity change after exposure). All experimental data were collected from pozzolan-amended cement samples (35 vol % pozzolan) exposed to brine saturated with 21 mole % H_2S and 79 mole % CO_2 for 2.5 days and 28 days.

The model is able to predict the experimentally observed (via SEM-BSE) formation of a calcite-rich layer after 2.5 days of exposure (Figure 2-a and Figure 2-b). For the 28-day exposure scenario, the predicted location and width of the calcite-rich layer did not match well need to the SEM-BSE analysis result (Figure 2-c and Figure 2-d).

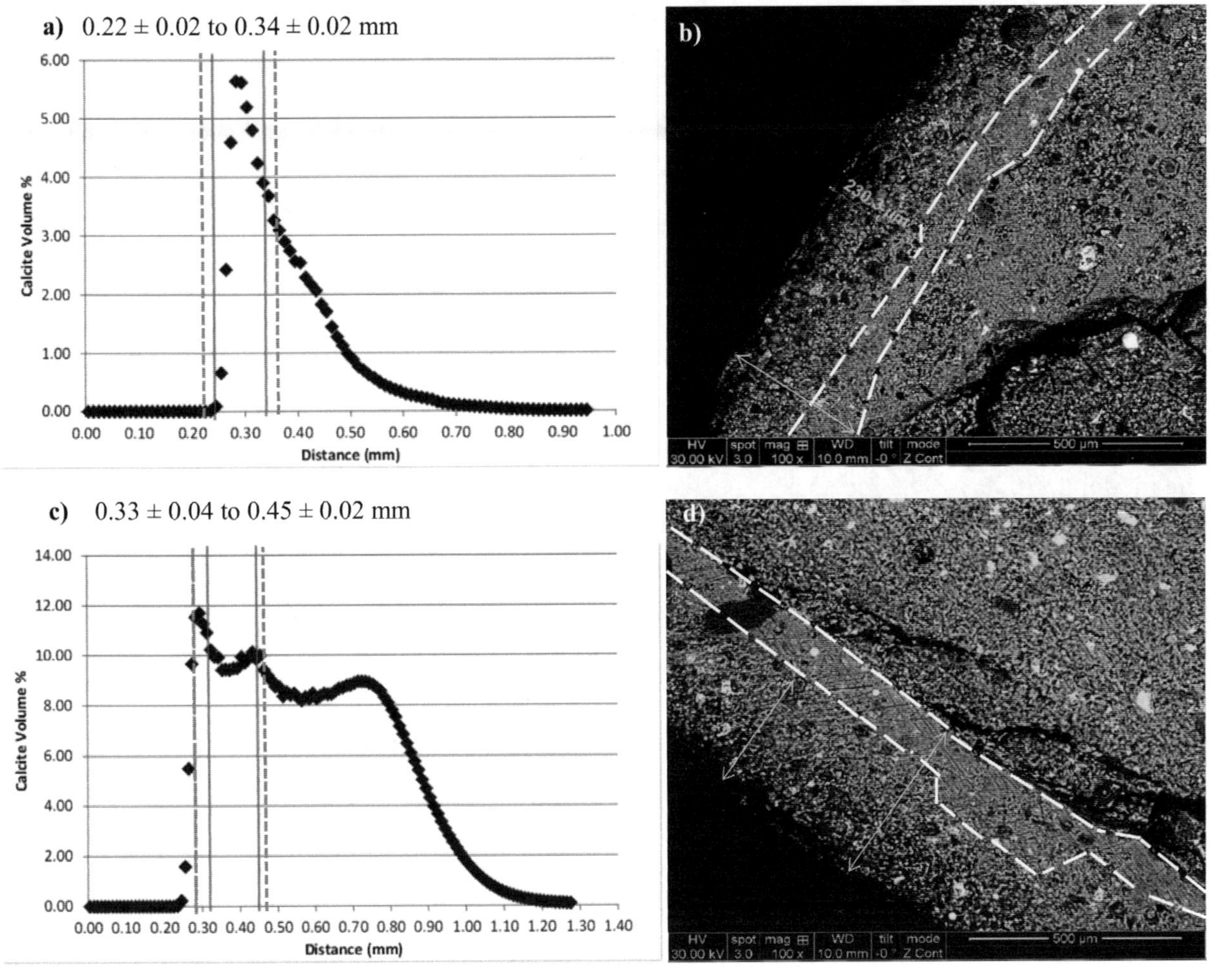

Figure 2: a) Model simulation result of calcite volume % as a function of distance into the cement from the brine-cement interface (0 mm is the interface). Exposure time: 2.5 days. The solid lines represent the width of the calcite-rich layer observed in SEM-BSE imaging and the dashed lines represent the uncertainty range of the layer. b) An example of an SEM-BSE image of 35 : 65 sample exposed for 2.5 days. The region in the middle of two white dashed lines is the calcite-rich layer. c) Model simulation result of calcite volume % as a function of distance into the cement from the brine-cement interface. Exposure time: 28 days. The solid lines and the dashed lines have the same meanings as those in 1-a. d) An example of an SEM-BSE image of 35 : 65 sample exposed for 28 days. The location of the calcite-rich layer in this image is from 0.30 mm to 0.43 mm.

The model predicts a 37 % increase in porosity at a narrow region very close to the interface (the inner frontier of the high-porosity region is about 0.45 mm away from the interface) from 0 day of exposure to 28 days of exposure. The high-porosity region is very narrow, and the overall alteration of the pozzolan-amended cement properties was not significant. The porosity at the core of the sample remains unchanged after 28 days of exposure. An increase of 50 % in CO_2

concentration or in H_2S concentration in brine (simulation results not shown) does not contribute to significant change in porosity profile after 28 days of exposure.

5. Conclusions

With the use of CrunchFlow, a reactive transport model is developed to simulate chemical interactions between pozzolan-amended wellbore cement and CO_2/H_2S under typical acid gas co-sequestration conditions. The model is able to predict the experimentally observed formation of a calcite-rich layer near the surface of exposed cement. The predicted location of the calcite-rich layer from model simulation for 2.5 days of exposure fits experimental SEM-BSE data well. The model predicts a 37% increase in porosity for a very narrow region close to the interface after 28 days of exposure, but the porosity at the core region of the sample exposed for 28 days remains close to the porosity of the unexposed sample. Overall, the simulation results reveal that the porosity change of a 35:65 pozzolan-amended sample is very limited after 28 days of exposure.

Acknowledgements

This work was supported by the Carbon Sequestration Program of the U. S. DOE National Energy Technology Laboratory (NETL) and U. S. DOE National Energy Technology Laboratory—Regional University Alliance (RUA) Partnership. The cement exposure experiments were conducted at EERC under the Plains CO_2 Reduction Partnership (PCOR), which is supported by NETL. Samples were analyzed at Carnegie Mellon. The authors also thank Ron Ripper at Civil and Environmental Engineering Lab of CMU, Tom Nuhfer and Jason Wolf at Material Science and Engineering Lab of CMU for assistance with various aspects of the analytical work.

References

Duan, Z.H., Sun, R., 2003. An improved model calculating CO2 solubility in pure water and aqueous NaCl solutions from 273 to 533 K and from 0 to 2000 bar. Chemical Geology, 193, 257–271.

Duan, Z.H., Sun, R., Liu,R., Zhu, C., 2007. Accurate thermodynamic model for the calculation of H_2S solubility in pure water and brines. Energy & Fuels, 21, 2056–2065.

Gasda, S. E., Bachu, S. and Celia, M. A., 2004. Spatial characterization of the location of potentially leaky wells penetrating a deep saline aquifer in a mature sedimentary basin. Environmental Geology, 46(6-7): 707-720.

Hawthorne, S.B., Miller, D.J., Holubnyak, Y., Harju, J.A., Kutchko, B.G., Strazisar,B.R., 2011. Experimental investigations of the effects of acid gas (H_2S/CO_2) exposure under geological sequestration conditions. Energy Procedia, 4: pp 5259–5266.

Jacquemet N., Pironon J. and Saint-Marc J., 2008. Mineralogical changes of a well cement in various H_2S-CO_2 (-brine) fluids at high pressure and temperature. Environ. Sci. Technol., 42 (1), 282–288.

Jacquemet N., Pironon J., Lagneau V., et al, 2012. Armoring of well cement in H_2S-CO_2 saturated brine by calcite coating—Experiments and numerical modeling. Applied Geochemistry, 27: 782-795.

Kutchko, B. G., Strazisar, B. R., Hawthorne, S. B., Lopano, C. L., Miller, D. J., Hakala, J. A., Guthrie, G. D., 2011. H_2S–CO_2 reaction with hydrated Class H well cement: Acid-gas injection and CO_2 Co-sequestration. International Journal of Greenhouse Gas Control, 5 (4), pp 880–888.

Steefel, C.I. and Lasaga, A.C., 1994. A coupled model for transport of multiple chemical species and kinetic precipitation/dissolution reactions with application to reactive flow in single phase hydrothermal systems. American Journal of Science, 294 (5): 529-592.

Steefel C. I., 2009. CrunchFlow User's Manual. Lawrence Berkeley National Laboratory, Berkeley, CA.

Theresa L., Watson T.L. and Bachu S., 2009. Evaluation of the Potential for Gas and CO_2 Leakage Along Wellbores. SPE Drilling & Completion, 24(1): 115-126.

(132e) Aspects of Fluid Compressibility, Swelling, and Time-Dependent Boundary Conditions for CO_2 Sequestration in Saline Aquifers

Philip Cheng, Department of Chemical and Environmental Engineering, Yale University, New Haven , CT and Abbas Firoozabadi, Department of Chemical and Environmental Engineering, Yale University, New Haven, CT

A number of theoretical studies have examined the onset and subsequent development of density- driven convective flow for carbon dioxide (CO_2) sequestration in saline aquifers. These studies employ a similar set of assumptions and boundary conditions, some of which may not be justified. For instance, the brine/water is taken to be incompressible (i.e., is described by a divergence-free velocity field), and the interface between the CO_2 and brine/water remains fixed in position and saturated at a constant CO_2 concentration. Our work involves a theoretical study of these specific aspects. We have performed a linear stability analysis to verify that fluid incompressibility may indeed be a good approximation. The stability analysis shows that fluid compressibility has only a small effect on the critical time and critical wavelength which characterize the onset of the density-driven convection. However, the assumption of a fixed interface may not be realistic. Dissolution of CO_2 into water can lead to swelling that may increase the volume of the aqueous phase by as much as seven percent. Furthermore, a fixed interfacial concentration can be achieved only if very large amounts of CO_2 are injected into the aquifer. Doing so can potentially lead to excess pressure buildup and CO_2 leakage from the aquifer. We have performed numerical simulations of experiments where CO_2 is mixed with water in diffusion-convection cells. Our model includes a moving interface where the time-varying CO_2 concentration is computed rigorously from an equation of state and thermodynamic principles. Incorporation of these features allow for closer agreement between our numerical results and experimental pressure evolution curves.

Results of Latest U.S. Department of Energy Funded Chemical Enhanced Oil Recovery Projects

S. A. Jikich, U.S Department of Energy/National Energy Technology Laboratory

This presentation reports on U.S. Department of Energy sponsored research leading to improved understanding of chemical methods for enhanced oil recovery (EOR) and mobility studies for cold, heavy oils reservoirs.

The improvement in mobility control of heavy oil examines polymer flooding and strategic use of gels to provide a more efficient sweep of North Slope heavy-oil reservoirs. The limits of applying polymer flood for heavy viscous oil were re-examined and redefined. New polymers for effective flooding of viscous oil were tested and new mechanisms for displacement by polymer solutions were proposed. Finally, novel methods for using polymer gels during recovery of unconventional viscous oils were examined.

Advancements in improved understanding of how chemical EOR works in laboratory and in the field include characterization and test of current and next generation of high performance surfactants for improved chemical flooding technology, reservoir characterization and geologic modeling role in implementation of Alkaline-Surfactant-Polymer (ASP) flooding technology and design of efficient chemical formulations for chemical flooding processes that are necessary for independent oil producers..

The results show that only a synergistic approach between chemical EOR, reservoir characterization, modeling, formation evaluation and reservoir management can lead to increased recovery factors.

Using water soluble polymers to recover viscous oil from Alaska's North Slope.

This project examined the potential of polymer flooding for recovering viscous oils when the polymer is able to reduce the residual oil saturation (Sor) to a value less than that of a waterflood. Experiments and fractional flow calculations were used to examine this question for conditions in North Slope reservoirs with viscous oils. Variables considered included oil viscosity, water/polymer viscosity, relative permeability characteristics, connate water and residual oil saturations, formation layering, presence/absence of cross flow, and pore volume throughput. It was found that induced changes in Sor can make a significant difference to recovery efficiency. As expected, the impact of Sor reduction by a polymer flood on oil recovery is more pronounced in reservoirs where residual oil saturations are high at the start of polymer flooding. The impact of Sor reduction diminishes with increasing degree of heterogeneity. A polymer flood can be effective for recovery of viscous oils even if the reservoir is extensively waterflooded before application of the polymer flood. A reduction in Sor was beneficial for all waterflood scenarios that were examined.

The project also examined the rheology of a new sulfonic associative polymer in porous media. For hydrophobically associative polymers, incorporating a small fraction of hydrophobic monomer into an HPAM polymer can promote intermolecular associations and thereby enhance viscosities and resistance factors. A tetrapolymer with low hydrophobic monomer content and a molecular weight (Mw) of 12-17 million g/mol. was compared with a conventional hydrolyzed polyacrylamide (HPAM) with 18-20 million g/mol molecular weight (Mw) and 35-40% anionic content. Rheological properties were similar for the two polymers. For both polymers in cores with permeabilities from 300-13,000 mD, no face plugging or internal filter cake formation was observed, and resistance factors correlated well using the capillary bundle parameter. For the HPAM polymer in these cores, low-flux resistance factors were consistent with low-shear-rate viscosities. In contrast, over the same permeability range, the associative polymer provided low-flux resistance factors that were 2-3 times the values expected from viscosities. Moderate shear degradation did not eliminate this effect—nor did flow through a few feet of porous rock. Propagation experiments in long cores (up to 5 ft) suggest that the unexpectedly high resistance factors could propagate deep into a reservoir—thereby providing enhanced displacement compared with conventional HPAM polymers. Compared with HPAM, the new polymer shows a significantly higher level of shear thinning at low fluxes and a lower degree of shear thickening at high fluxes. The associative polymer appears to contain a species that propagates through porous rock at rates comparable to those for HPAM and a second species that moves much slower and creates much higher resistance factors. This may improve recovery but may also present problems with injectivity and polymer transport.

The project also revealed that in-depth profile modification is most appropriate for high permeability contrasts (e.g. 10:1), high thickness ratios (e.g., less-permeable zones being 9 times thicker than high-permeability zones), and relatively low oil viscosities. Because of the high cost of the blocking agent relative to conventional polymers (gel polymer concentration is 5-10 higher than that of conventional polymer flooding), economics favor small blocking-agent bank sizes (e.g. 5% of the pore volume in the high-permeability layer). Even though short-term economics may favor in-depth profile modification, ultimate recovery may be considerably less than from a traditional polymer flood.

Mechanical degradation of an associative polymer versus HPAM was performed to investigate the effects of reservoir brine salinity and divalent cations on mechanical (shear) degradation of a new hydrophobic associative polymer, (SNF C1205). This polymer has a small fraction of hydrophobic monomer groups along the acrylamide/acrylate backbone. Degradation was induced by forcing polymer solutions through a short core plug using a wide range of rates (1 to 500 ft/d). The study investigated solutions with salinities from 0.252 to 12.65% total dissolved solids and with divalent cation (Ca2+) concentrations from 0 to 1.15%. Effluent viscosities and screen-factors were measured to determine the degree of mechanical degradation. For comparison, a conventional polyacrylamide (HPAM) polymer with similar molecular weight (18-20 million g/mol) was submitted to the same tests. As expected, mechanical degradation of the associative polymer increased with velocity in porous media, and rheology of the associative polymer in the viscometer showed Newtonian behavior at low shear rates and shear thinning at higher shear rates. As with HPAM, calcium exerted a stronger influence than sodium in reducing solution viscosity. However, tests demonstrated that the degree of mechanical degradation of associative polymer solutions does not always increase with ionic strength. This finding differs from previous reports for HPAM. When compared with the conventional HPAM, even though the associative polymer may show more shear sensitivity under some circumstances, it still provided higher viscosity and screen factor, and presumably, better mobility control than the conventional HPAM.

Implementation of Alkaline-Surfactant-Polymer Flood in Illinois Basin

Reservoir characterization studies presented in this work are supporting pilot and field wide expansion of Alkaline-Surfactant-Polymer Floods (ASP) in two distinct sandstone reservoirs in Lawrence Field, Lawrence County, Illinois. Lawrence Field has produced in excess of 410 million barrels of oil since 1906 through primary and secondary waterflood techniques. The field currently produces at the rate of less than 2% oil cut and, like many mature fields in the United States, is approaching its economic limit even though there is much remaining recoverable oil. Using ASP flood test results, it is estimated that 130 million barrels of oil can be recovered using ASP in reservoirs in Lawrence Field. Alkaline-Surfactant-Polymer (ASP) flooding is an untested technology in the Illinois Basin and this project presents an exceptional opportunity to perform and document field testing of this unconventional recovery technique on two of Illinois's most prolific reservoirs.

Rex Energy has committed resources to implement an ASP flood in Lawrence Field. The research conducted by the Illinois State Geological Survey is intended to optimize the opportunity for success of this project.

Most of the unrecovered oil in Lawrence Field is contained in Pennsylvanian Age Bridgeport sandstones and Mississippian Age Cypress sandstones. These reservoirs are highly complex and compartmentalized. Detailed reservoir characterization including the development of 3-D geologic and stochastic models of target areas in the field is needed to identify areas with the best potential to recover remaining reserves including unswept and by-passed oil.

This four year project was designed to compile, interpret, and analyze the data required to conduct reservoir characterization for the Bridgeport and Cypress sandstones in pilot areas and expansion areas for a field wide implementation of ASP flooding in Lawrence Field. Geologic and stochastic modeling needed for reservoir characterization and reservoir simulation were completed as prerequisites for design of efficient ASP flood patterns. Characterizing the complex reservoir geology that identifies the geologic conditions that will optimize oil recoveries for expansion of the ASP flood from small, limited pilots in the Bridgeport and Cypress sandstones to the remainder of Lawrence Field is the primary objective of this project. It will permit evaluation of efficiency of oil recovery from Bridgeport and Cypress sandstone reservoirs using ASP technology. Additional objectives of this project include assessing the applicability of ASP flood technology to similar reservoirs in the Illinois Basin and other U.S. reservoirs.

Designing chemical floods for Kansas reservoirs

The principle objective of the project was to design efficient chemical formulations for chemical flooding processes that are necessary for independent oil producers to make an informed assessment for implementation of a pilot or demonstration project.
Ten Kansas oil reservoirs/leases were studied through geological and engineering analysis to assess the potential performance of chemical flooding to recover oil. Reservoirs/leases that have been efficiently waterflooded have the highest performance potential for chemical flooding. Laboratory work to identify efficient chemical systems and to test the oil recovery performance of the systems was the major effort of the project. Efficient chemical systems were identified for crude oils from nine of the reservoirs/leases. Oil recovery performance of the identified chemical systems in Berea sandstone rocks showed 90+ % recoveries of waterflood residual oil for seven crude oils. Oil recoveries increased with the amount of chemical injected. Recoveries were less in Indiana limestone cores. One formulation recovered 80% of the tertiary oil in the limestone rock. Geological studies for nine of the oil reservoirs are presented. Pleasant Prairie, Trembley, Vinland and Stewart Oilfields in Kansas were the most favorable of the studied reservoirs for a pilot chemical flood from geological considerations. Computer simulations of the performance of a laboratory coreflood were used to predict a field application of chemical flooding for the Trembley Oilfield. Estimates of field applications indicated chemical flooding is an economically viable technology for oil recovery.

Next Generation Surfactants for Improved Chemical Flooding Technology

The objective of this project was to characterize and test current and next generation high performance surfactants for improved chemical flooding technology, focused on reservoirs in the Pennsylvanian-aged (Penn) sands in Oklahoma. Recent models relating surfactant structure were extended to optimization of microemulsion formulation through the relationship between the theoretically accessible packing factor (Pf) calculation and the experimentally measurable characteristic curvature (Cc) of the surfactant in the surfactant membrane in a microemulsion. This was accomplished by extending the range of surfactants for which the Cc has been measured to surfactants typical of those used for EOR. This included the study of a suite of novel petroleum sulfonates, the recently commercialized extended surfactants (alkylethoxypropoxy sulfates), and a new series of disulfonates optimized for microemulsion formation. In particular, the current Pf calculation was modified to better account for multiple and branched hydrophobes, which improved the correlation between the calculated Pf and the measured Cc.
The relationship between new surfactants, co-surfactant (e.g., alcohol) concentrations, pH, and the use of sacrificial agents to minimize the surfactant adsorption at the brine/rock interface were studied. While current surfactant concentrations for EOR are much lower than those used earlier, the ability to reduce those concentrations still further is limited by the high adsorption of the surfactant by the reservoir rock. Improving the ability to design low interfacial tension (IFT) formulations that simultaneously have ultralow adsorptions on the reservoir rock can be a significant advance in improving the commercial viability of surfactant flooding technology.

Correlations between the optimal salinity, the Cc of the surfactant, the critical microemulsion concentration (CμC), and the achievement of ultralow interfacial tensions at concentrations below the CμC was also researched. It has been shown that ultralow IFT can be achieved without the formation of a microemulsion. If a correlation between ultralow IFT in the absence of microemulsion formation and the microemulsion phase diagram can be determined, then it will be possible to use recent design equations for microemulsion formulation to design surfactants to produce ultralow IFTs at concentrations below those required for the formation of a middle phase (Winsor Type III) microemulsion.

The University of Oklahoma, in partnership with a small independent producer, demonstrated the use of a novel surfactant formulation on two high-water-cut stripper wells—wells that produce less than 10 barrels of oil per day and produce significantly more water than oil—located in the Bartlesville formation in northeast Oklahoma. In the first well, the chemical flood with surfactant concentration of 0.5wt% mobilized 75 percent of the residual crude oil. In the second well, with about half the total surfactant concentration of the first well (0.23 wt%ent), more oil was mobilized (approximately 80 percent of the residual crude oil).

Based on these results, two pilot projects in Northeast Oklahoma are being designed for implementation. It is estimated that widespread application of these techniques could result in economic production of an additional billion barrels of oil from the Bartlesville formation.

(229a) High-Performance Supercapacitor and Li Ion Battery Electrodes Using Novel Ordered Mesoporous Metal Oxide Materials

Jinwoo Lee, Department of Chemical Engineering and School of Environmental Science & Engineering, Pohang University of Science and Engineering, Pohang, South Korea

For high performance supercapacitor and Li-ion battery electrodes, various mesoporous metal oxide materials have been synthesized by employing lab-synthesized block copolymers as a structure-directing agent or ordered mesoporous silica materials as hard templates. An ordered mesoporous WO_{3-X} material with high electrical conductivity was employed for use as a supercapacitor electrode in 2.0 M H_2SO_4. The electrode made of mesoporous WO_{3-X} exhibited a high rate capability and an excellent capacitance (366 $mFcm^{-2}$, 639 Fcm^{-3}), which were attributed to the large ordered mesopores, high electrical conductivity, and high material density. Fe_3O_4 nanocrystals confined in mesocellular carbon foam (MSU-F-C) are synthesized by '*Host-Guest*' approach and tested as anode for lithium-ion batteries (LIBs). 13-27 nm magnetite Fe_3O_4 nanocrystals are formed inside ~30 nm pores of mesocellular carbon matrix. Fe_3O_4/MSU-F-C (45 wt% Fe_3O_4) exhibited 1007 mA h g^{-1} of first charge (Li de-insertion) capacity at 0.1 A g^{-1} and 111% of capacity retention at 150th cycle, and retains 37% capacity at 7 A g^{-1} (~7 C rate). For the first time, we synthesized mesostructured LTO/carbon composite (Lithium titanate/carbon) with large (> 15 nm) and uniform pores for high performance LIB anodes via a block copolymer assembly. The electrochemical characterization of this material will be presented.

(229b) 3-D Graphenic Network Composites for High-Performance Li-Ion Battery Electrodes

Xin Zhao[1], Cary M. Hayner[1], Kai Han[1], Jung Kyoo Lee[2], Mayfair C. Kung[1] and **Harold H. Kung**[1], (1)Chemical and Biological Engineering, Northwestern University, Evanston, IL, (2)Chemical Engineering, Dong-A University, Pusan, South Korea

Three-dimensional, highly conducting, networks constructed from graphene sheets have been found to be very useful scaffolds to form composite materials with high charge-storage capacities for Li ion batteries. Additional treatment of the graphene sheets to introduce in-plane defects greatly enhances the cross-plane ionic conductivity, which results in much enhanced rate capability of such materials. Such materials have exhibited power density as high as that of supercapacitors. We have constructed electrode materials suitable for use as anode or cathode for Li ion batteries. For anode materials, a three-dimensional graphenic network with in-plane defects exhibits a stable, repeatable power capability of 180 mAh/g at a rate of 2 A/g, corresponding to a recharge rate of less than 5 min. Imbedding Si nanoparticles into the network greatly improves the charge storage capacity, achieving >3100 mAh/g at low rates, and >1000 mAh/g at a high rate of 8 A/g, with a coulomb efficiency as high as 99.9%, and typically >99.6%. Cathode materials can be formed by imbedding metal fluorides or sulfides/sulfur into these scaffolds. These materials also exhibit a much higher charge storage capacity than the conventional Li metal oxides (M= Co, Ni, Mn). However, capacity fade remains an issue that needs to be overcome. The characterization results of these materials will be presented.

(229c) High Energy Density Semiconductor-Carbon Nanotube Anodes for Lithium Ion Batteries

Brian J. Landi[1], Roberta A. DiLeo[2], Melissa Thone[3], Michael W. Forney[2], Alan Raisanen[4], Matthew J. Ganter[5], Jason Staub[2] and Reginald E. Rogers Jr.[1], (1)Chemical & Biomedical Engineering, Rochester Institute of Technology, Rochester, NY, (2)NanoPower Research Laboratories, Rochester Institute of Technology, Rochester, NY, (3)Chemical & Biomedical Engineering, Rochester Institute of Technology, (4)KGCOE Semiconductor Microsystems Fabrication Laboratory, Rochester Institute of Technology, (5)Golisano Institute for Sustainability, NanoPower Research Laboratories, Rochester Institute of Technology, Rochester, NY

Alternative active materials for lithium ion batteries are under investigation to meet increasing energy storage demands. Ultra high capacity Si and Ge are promising anode materials that demonstrate capacities much higher than state-of-the-art graphite. In addition, promising developments have been made recently with electrode designs employing carbon nanotubes (CNTs) as a conductive additive or a free-standing electrode (absent of any binder or metal current collector) to serve as a physical support for silicon or germanium increasing energy and power densities. Several approaches have shown viability such as semiconductor film deposition using chemical vapor deposition (CVD) and a nanoscale hybridapproach. The performance of Si and single wall CNTs (SWCNT s) anodes, fabricated by low pressure chemical vapor deposition (LPCVD) and plasma enhanced chemical vapor deposition (PECVD), was measured to have anode extraction capacities between 1000-1500 mAh/g, even at Si weight loadings around 30%. However, dramatic differences in the rate capability and cycling for anodes from these two deposition techniques were demonstrated. Alternatively, the mixing of Ge nanoparticles (NPs) with SWCNTs to form a hybrid free-standing anode has enabled a tunable 3-dimensional electrode structure. These hybrid anodes exhibit capacities over 800 mAh/g at rates up to 1C. The balance between rate and capacity using Si and Ge has led to the development of a mixed hybrid anode that contains both materials in concert with a CNT electrode. The addition of a Si coating enhances the Ge NP anode capacity along with reducing the surface area and the first cycle loss. Full batteries incorporating hybrid anodes and conventional cathodes show potential improvements in gravimetric energy density over state-of-the-art by more than 50%.

(229d) Electrospun Nanofibers for Lithium Ion Batteries

Nathaniel S. Hansen[1,2], Daehwan Cho[2], Kyoung Woo Kim[2], Jay Hoon Park[2], Brian Williams[2] and Yong Lak Joo[2], (1)Axium Nanofibers, Ithaca, NY, (2)School of Chemical and Biomolecular Engineering, Cornell University, Ithaca, NY

Electrospinning is now a relatively common method for the production of one-dimensional nanomaterials from a scalable, cost-effective process. Recently, we have developed an approach to generate inorganic nanofibers with various nanostructures through a high-throughput, water-based solution process. We have utilized this process to generate nanofibers for lithium ion battery anode and cathode structures. It is well understood that the utilization of silicon at the anode requires nano-synthesis to overcome pulverization, and thus we have generated carbon/silicon composite anode structures that display an extremely high energy capacity (2100 mAh/g) over many cycles. Further, we display control over the carbon/silicon composite structure to minimize capacity loss for long lifetime performance. However, it is known that the generation of high energy density anode materials will not drastically improve the energy density of the battery as a whole until a high energy density cathode is synthesized. To this end, we have synthesized similar cathode materials consisting of carbon/lithium sulfide composite structures that similarly display high energy density and long lifetime. These materials, when paired together, are shown to have one of the highest theoretical energy densities thus enabling us to cost-effectively produce the highest energy density battery as a whole.

(229e) Cobalt Oxide, Multi-Wall Carbon Nanotube Polymer Network for Lithium Ion Battery Anodes

Joshua Hill[1], Yun Xu[2], Ling Fei[2], Yuling Li[1] and Hongmei Luo[3], (1)Chemical Engineering, New Mexico State University, Las Cruces, NM, (2)Chemical Engineering Department, New Mexico State University, Las cruces, NM, (3)Chemical Engineering, New Mexico State University, Las cruces, NM

Much research has been focused on the field of energy storage in lithium ion batteries. A family of cathode materials consisting of a polymer network containing cobalt oxide, and carbon nanotubes has been prepared. This combination of the polymer network with nanocrystalline cobalt oxide and multi-wall carbon nanotubes provides a novel carbon framework capable of shuttling and storage of lithium ions. Colloidal cobalt oxide was first prepared by the thermal decomposition of a metal-organic precursor. The multiwall carbon nanotubes were in used in conjunction during the synthesis of the colloidal cobalt oxide. The polymer network was prepared by mixing nanocrystalline cobalt oxide, multiwall carbon nanotube solution, with poly(ethylene)imine solution and ethylenediaminetetraacetic acid, followed by calcination at temperatures ranging from 400°C to 700°C. Characterization of the anode material was accomplished using X-ray diffraction, and scanning electron microscopy. Electrochemical properties of the materials were tested in lithium ion coin cell batteries constructed in an inert atmosphere.

(229f) Hybrid Tin Oxide Nanotube Arrays As Anodes for Lithium Ion Batteries

Umang V. Desai and Di Gao, Chemical and Petroleum Engineering, University of Pittsburgh, Pittsburgh, PA

Compared to traditional anodes based on nanostructured powdered paste, directly grown, one-dimensional ordered nanostructured array offers remarkable advantages such as fast-electron transport/collection and ion diffusion, sufficient electrochemical reaction of individual nanostructures, enhanced material-electrolyte contact area and facile accommodation of the strains caused by lithium intercalation and de-intercalation. Amongst various ordered metal oxide arrays, SnO_2 has been regarded as one of the most promising candidates for alternative anode material for Li-ion batteries due to the high theoretical capacity (781 mAh/g) compared to that of graphite (372 mAh/g). Thus, making 1D oriented nanostructure array of SnO_2 on metal substrate over a large area and further applying it as anode of battery is highly desirable for the exploration of novel electrode materials. Large-scale growth of uniform SnO_2 nanotube arrays on Ti substrate, with high orientation consistency, was achieved through a low temperature hydrothermal process. Furthermore, hybrid Sn-SnO_2 nanotubes were created by reducing the as-synthesized SnO_2 nanotubes. The performance of these arrays as anode materials in lithium-ion batteries was evaluated.

(229g) Facile Fabrication of Hollow TiO2 Nanoparticles As Safe Anode Materials for Lithium Ion Batteries

Jian Zhu, Chemical Engineering, Wayne State University, Detroit, MI, K. Y. Simon Ng, Department of Chemical Engineering and Materials Science, Wayne State University, Detroit, MI and Da Deng, Chemical Engineering and Materials Science, Wayne State University, Detroit, MI

Lithium ion batteries will play increasingly important role in expanding market of electric vehicles and green energy storage, in addition to continuously dominant role in personal electronic devices. The recall of lithium ion batteries for laptops and Chevy Volt battery fires demonstrate the critical importance of battery safety, especially for large-scale production/application in future. Safety concern requires next-generation lithium ion batteries to have carbon-alternative anode materials where its voltage plateau is higher than that of carbon to enhance safety without compromising cyclability. TiO_2 could be potentially used as such safer anode materials for lithium ion batteries, and its structural stability guarantees good cyclability in reversible lithium ion storage. In this presentation, our recent efforts on facile preparation of hollow TiO_2 nanoparticles will be reported. The hollow porous nanoparticles are formed by aggregation of smaller nanoparticles as basic building units. The hollow TiO_2 nanoparticles with high specific surface area and short diffusion path demonstrate excellent cycling performance and high current rate in reversible lithium ion storage at significantly higher voltage plateau than that of carbon.

(296a) A Nanoscale Study of the Formation of the Solid Electrolyte Interphase in Li Ion Batteries

Yanjing Chen[1,2], Mengyun Nie[3], Brett Lucht[3] and Arijit Bose[1,2], (1)Chemical Engineering, University of Rhode Island, Kingston, RI, (2)Rhode Island Consortium for Nanotechnology and Nanoscience, Kingston, RI, (3)Chemistry, University of Rhode Island, Kingston, RI

The Solid Electrolyte Interphase (SEI) is a passivating film formed on the surface of graphite anodes because of the reductive decomposition of electrolyte solvents during initial charging of a lithium ion battery (LIB). The SEI plays a crucial role on the cycling performance and storage life of LIB, protecting the electrolyte from complete decomposition but also providing additional resistance to Li ion transport. However, a comprehensive understanding of the structure and composition of the SEI is still lacking.

We have developed an in-house transmission electron microscope (TEM grid) embedded graphite/Li half cell, with a binder-free (BF) graphite anode prepared using electrophoretic deposition (EPD) of graphite nanoparticles. Using such a cell removes the 'noise' created by inactive components such as binders, and allows us to focus directly on the role of different electrolytes on the formation of the SEI and its relationship to the cell performance, and the growth of SEI at different voltages during initial charging. We were able to follow the growth of the SEI *in-situ* at the nanoscale. Coupled with X-ray photoelectron spectroscopy (XPS), Fourier transform infrared spectroscopy (FTIR) and Nuclear magnetic resonance (NMR), we have been able to identify the morphology and key chemical components of the SEI and how these vary with different electrolytes. Our results provide crucial insights into optimizing choices of electrolytes and anode materials for Li-ion batteries.

(296b) Highly Ordered Mesoporous Materials As Potential Electrodes for Li-Ion Batteries

Feng Jiao, Bryan Yonemoto and Gregory Hutchings, Chemical Engineering, University of Delaware, Newark, DE

Highly Ordered Mesoporous Materials as Potential Electrodes for Li-ion Batteries

Bryan Yonemoto, Greg Hutchings, Feng Jiao*

Center for Catalytic Science & Technology, Department of Chemical Engineering, University of Delaware, Newark, DE, 19716 USA. Fax: +1-302-831-1048; Tel: +1-302-831-3679; E-mail: jiao@udel.edu

A major goal of lithium battery research is to prepare electrodes that can sustain higher rates than present materials, for new high power applications such as HEV's. Nanomaterials support high rates, although often at the expense of volumetric energy density. Nanostructured materials can offer advantages compared with nanoparticles. For example, we have shown that mesoporous β-MnO_2 is capable to accommodate 0.92 Li^+ per Mn, while in a sharp contrast, Li^+ ion cannot intercalate into micron sized β-MnO_2. Here, we will show our recent progress in the synthesis of novel nanostructured materials as potential electrode materials for lithium-ion batteries.

There is much current interest in mesoporous metal oxides. The synthesis of such solids, especially porous materials, often requires a templating approach. Soft templates, e.g. long surfactant molecules, or hard templates, e.g. mesoporous silica, may be used. However, it is difficult to prepare metal oxides with extremely large pores (> 10 nm) by such routes. Overcoming this challenge is important if we are to prepare mesoporous transition metal oxides with enhanced electrochemical properties. We will describe the synthesis of a wide range of mesoporous transition metal oxides by using mesoporous silica as hard template. The as-prepared mesoporous compounds exhibit superior performance compared with their normal analogues.

References:

1. Ren, Y., Armstrong, A. R., Jiao, F. & Bruce, P. G. Influence of Size on the Rate of Mesoporous Electrodes for Lithium Batteries. Journal of the American Chemical Society 132, 996-1004 (2010).

2. Jiao, F., Bao, J. L., Hill, A. H. & Bruce, P. G. Synthesis of Ordered Mesoporous Li-Mn-O Spinel as a Positive Electrode for Rechargeable Lithium Batteries. Angewandte Chemie-International Edition 47, 9711-9716 (2008).

(296c) V2O5 Network Structure As Cathode for Lithium Ion Batteries

Yun Xu, Chemical Engineering Department, New Mexico State University, Las cruces, NM, Marco Dunwell, Chemical Engineering, New Mexico State University, Las Cruces, NM and **Hongmei Luo**, Chemical Engineering, New Mexico State University, Las cruces, NM

V_2O_5 has been vigorously investigated as an electrode material because of its high output voltage, and low cost, and high capacity compared to most commercial cathode materials, including $LiFePO_4$, $LiMn_2O_4$, and $LiCoO_2$. The unique crystal structure which is formed by the stacking of V_2O_5 layers makes V_2O_5 a special host for the reversible insertion and extraction of lithium ions. Many works has been devoted to improve the performance of V_2O_5, including making V_2O_5/carbon matrix composite, V_2O_5 porous structure, and nanostructures, etc.

In this work, we report a V_2O_5 network structure which is synthesized by a polymer-assisted chemical solution method. This V_2O_5 network structure delivers high and stable capacity of around 300, 200, and 120 mAh/g at current densities of 100, 400, and 800 mA/g after 50 cycles, respectively, which demonstrated an excellent stability and rate capability. The capacity and rate capability is better than most reported non-carbonaceous V_2O_5 structure. The high performance of the network structure comes from the unique two-dimensional network which bridges different length scales from the nanoscale to the micro/macro scale.

(296d) Polyaniline Nanofiber/Vanadium Pentoxide Layer-by-Layer Electrodes for Energy Storage

Lin Shao, Chemical & Environmental Engineering, Yale University, New Haven, CT, Ju-Won Jeon, Texas A&M University, College Station, TX and Jodie Lutkenhaus, Department of Chemical Engineering, Texas A&M University, College Station, TX

Li-ion batteries have garnered much interest in portable power, but in some aspects, they are limited. For example, some commercial Lithium-ion battery electrodes are subject to volumetric expansion, safety concerns, and high cost. However, nanostructured electrodes comprised of alternative materials can potentially address many of these limitations. Here, we report the layer-by-layer assembly of polyaniline (PANI) nanofibers and V_2O_5 to form hybrid electrodes for electrochemical energy storage. PANI nanofibers assist in forming a highly porous electrode (density = 0.247g/cm^3), and impart electronic conductivity. Both PANI and V_2O_5 store charge through Faradaic reactions. Together, this composite may form an electrode that surpasses either material alone. The growth of the film was investigated using profilometry and quart crystal microbalance; the thickness of these films increased linearly with respect to the number of layer pairs deposited. The non-aqueous electrochemistry of the electrode was explored using lithium as a counter and reference electrode. It was found that the electrochemical response possessed contributions from both PANI nanofibers and V_2O_5, and that the electrochemical performance of this LbL system was dependant on film thickness. A maximum capacity of 233 mAh/g at a current of 2μA/cm^2 was measured for a film of 16 layer pairs. In addition, PANI nanofibers were found to dominate the electrochromic response measured using UV-Vis spectroscopy. The composition of the film and the oxidation state of vanadium were measured and monitored using X-ray photoelectron spectroscopy (XPS). No appreciable volumetric expansion after charge/discharge cycling. Based on our current study, the formation of pernigraniline base polyaniline was attributed to the performance degradation of these films. This study indicates how LbL assembly can be used to fabricate high-performance electrodes for electrochemical energy storage.

(296e) Engineering Lixalysizo Ionic Conductive Thin Films by Atomic Layer Deposition for Lithium-Ion Battery Applications

Ya-Chuan Perng[1], **Jea Cho**[1], Daniel Membreno[2], Nick Cirigliano[2], Bruce Dunn[2] and Jane P. Chang[1], (1)Department of Chemical and Biomolecular Engineering, University of California, Los Angeles, Los Angeles, CA, (2)Department of Materials Science and Engineering, University of California, Los Angeles, Los Angeles, CA

Lithium (Li)-ion batteries have drawn much attention for their outstanding performance in portable electronics applications with the potentials to function as a power source for further minizaturized devices, including micro-systems through the utilization of 3-dimensional electrodes based on high aspect ratio pillars. To realize this potential, an ultra-thin and highly conformal electrolyte layer is needed to coat the 3D electrode array. The ionic conductor lithium aluminosilicate ($LiAlSiO_4$) synthesized by atomic layer deposition (ALD) is a promising electrolyte material for 3D battery applications not only due to its high ionic conductivity along its c-axis resulting from channels formed by the alternating tetrahedra of aluminum-oxygen (Al-O) and silicon-oxygen (Si-O), but also expected to provide similar improved cell cyclability, as reported in the preliminary studies of ultra-thin metal-oxide ALD coatings on electrodes.

The self-limiting characteristic of ALD allows for precise control of thickness and composition of complex oxides and results in a highly conformal and pinhole-free coating suitable in 3D micro-battery applications or electrolyte surface coatings. The metal precursors used in this work are tetraethyl orthosilicate (TEOS), trimethylaluminum (TMA) and lithium t-butoxide (LTB). These precursors, along with water vapor as the oxidant, were used to deposit SiO_2, Al_2O_3 and Li_2O, with the deposition rates in the range of 0.8~2Å/cycle, respectively. The deposition rate of stoichiometric $LiAlSiO_4$ was ~20Å/cycle at a temperature of 290°C. The concentration of each metal element in $Li_xAl_ySi_zO$ (LASO) thin films was found to correlate closely to ALD cycles and the associated incubation times. The crystallinity of the films after post-deposition rapid thermal annealing (RTA) was a function of cation atomic percentage. Li-ionic conductivities and the activation energy of as-deposited LASO films with respect to lithium contents as well as their relation to the film thickness were studied. The LASO ALD coating on 3D features, such as NWs and nanopaticles (NPs), were confirmed to be conformal and uniform by transmission electron microscopy (TEM) imaging. The cell performance as well as cyclability enhancement from LixAlySizO was investigated for a silicon-nanowire 3D microbattery, where SiNW was used as an anode, to explore the potentials of a solid-state SiNW battery with a solid-oxide electrolyte.

(296f) Mixed Hydrophobic Room Temperature Ionic Liquid Electrolytes for Rechargeable Lithium-Air Battery

Mahbuba Ara[1], Lixin Wang[2], Kapila Wadumesthrige[2], Steven O. Salley[1] and K. Y. Simon Ng[3], (1)Chemical Engineering and Materials Science, Wayne State University, Detroit, MI, (2)Chemical Engineering, Wayne State University, Detroit, MI, (3)Department of Chemical Engineering and Materials Science, Wayne State University, Detroit, MI

The increase in solution resistance due to the evaporation and decomposition of carbonate based electrolyte especially at high charge voltages hinders the performance (cyclability) of Lithium –Air battery. Hydrophobic room temperature ionic liquid can be used to increase the cycle numbers of lithium-air battery by eliminating the decomposition of electrolyte as well as protecting the lithium anode from hydrolysis reaction with moisture from air. Pyrrolidinium based ionic liquid N-butyl-N-methylpyrrolidinium bis (trifluoromethanesulfonyl) imide ($PYR_{14}TFSI$) shows a wide electrochemical voltage stability window. But it has high viscosity and consequently lower conductivity than standard organic electrolyte. Therefore, the mixed ionic liquid that simultaneously has the properties of high ionic conductivity and high electrochemical voltage window for battery application has been investigated. Imidazolium based ionic liquid, 1-ethyl-1-methylimidazolium bis (trifluormethylsulfonyl) imide (EMIMTFSI) has lower viscosity and higher ionic conductivity than $PYR_{14}TFSI$. This study will investigate the addition of different proportions of EMIMTFSI to $PYR_{14}TFSI$ for higher ionic conductivity as well as high electrochemical voltage stability. Optimum mixture of the ionic liquids will be studied for the long term cyclic performances of lithium-air battery. Different concentrations of lithium salt (Lithium bis (trifluoromethanesulfonyl) imide (LiTFSI)) to the optimum mixture will be examined for the best performance of lithium-air battery. The optimum ionic liquid electrolyte will be used in conjunction with perovskite type bifunctional catalyst for enhanced capacity and rechargeability of lithium-air battery.

(296g) Novel Graphene Supported Bifunctional Catalysts Combined with Ionic Liquid Electrolyte in Rechargeable Lithium-Air Batteries for Prolong Cyclic Stability

Lixin Wang[1], Mahbuba Ara[2], Steven O. Salley[2] and K. Y. Simon Ng[3], (1)Chemical Engineering, Wayne State University, Detroit, MI, (2)Chemical Engineering and Materials Science, Wayne State University, Detroit, MI, (3)Department of Chemical Engineering and Material Science, Wayne State University, Detroit, MI

Li-air battery is widely considered to be the next generation battery chemistry with extremely high energy density comparable to gasoline. They are very promising technology to prolong the driving ranges of electric vehicles. Compare to their high theoretical value, the practical energy density of Li-air battery is much lower due to the slow kinetics of both discharge and charge reactions and hence, high overpotential at air cathode. To address this technical barrier, Pt catalyst has been used to overcome the activation energy and consequently reduce the overpotential. However, the high cost of Pt still restricts the wide application of Pt in air-cathode. In this presentation, low cost, bi-functional catalyst has been developed to replace Pt catalyst. A Li-Air system capable of 100 cycles with an average capacity of 2000 mAh/g and total energy efficiency of 75% has been achieved compared to 30 cycles with 1800 mAh/g average energy density and 85% energy efficiency as a control.

Another important factor that hinders the cycle life of Li-air battery is the increasing solution resistance due to the consumption of carbonate based electrolyte, especially during the high charge voltages. Electrolyte oxidation, evaporation and side reactions with active electrode materials are believed to be the main reason of the electrolyte lost. In this presentation, the conductivity, voltage window, viscosity and other electrochemical related properties of several ionic liquids as well as their mixture at different mass ration have been also investigated. by employing the optimized ionic liquid electrolyte, the high resistance after cycling is expected to be addressed and the cycle life of Li-air batteries are expected to be prolonged to 300 cycles.

"Fuel Processing: Renaissance of the Second Decade"

Brant A. Peppley, Rajesh Parmar and Kunal Karan
Queen's-RMC Fuel Cell Research Centre
Queen's University
Kingston Canada

The early development of fuel processing technologies for fuel cell systems was, to a great extent, driven by fuel cell automobile programs. In the early 1990s, an on-board reformer was believed to be the best option for fuel cell automobile commercialisation. Regrettably, the belief that we simply needed to adapt design concepts from large-scale industrial steam reforming hampered progress at the time.

Figure 1: 100MM SCFD reformer and waste heat recovery section supplied by Selas-Linde Fluid in a hydrogen plant designed, supplied, and constructed by Linde. Source: http://www.selas-linde.com/range_of_services/modular_constructions.php

Very early attempts to modify reformer designs to address the need for rapid load following and better heat transfer lacked the fundamental understanding of reactor design and reaction kinetics.

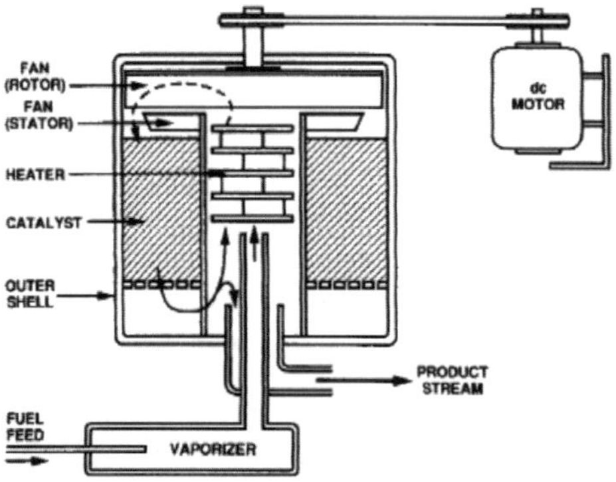

Figure 2: Schematic cross-section of a recirculating methanol steam reformer with enhanced heat transfer but very undesirable flow characteristics not to mention undesirable parasitic power requirements. (R.A. Lemons, J. Power Sources, 29 (1990) 251 – 264.

Autothermal (ATR) and Partial Oxidation (POX) reformers provided a partial solution to the heat transfer problems but resulted in a severe penalty in overall system efficiency. The idea of on-board reforming for automobiles was all but completely abandoned in the late-1990s and, the level of effort on fuel processing for fuel cell systems decreased drastically. A significant amount of progress, however, had been made prior to the reduction in support for fuel processor research and reformer designs based on structured catalysts and catalyst coated micro-structured heat transfer components had made significant progress.

Recent developments in stationary and portable fuel cell system products using hydrocarbons, alcohol and oxygenated fuels including diesel fuel, dimethyl ether, ethanol, methanol and propane as well as natural gas have created a renewed interest in developing improved fuel processors. Although these types of systems do not necessarily require fast start up and load following, significant improvements in overall efficiency, reductions in pressure drop through the bed and improved dynamic response can be and have been achieved. A better understanding of the coupling of heat transfer, catalytic kinetics and mass transfer limitations in conjunction with the discovery of new classes of catalysts that meet the demanding requirements of hydrocarbon fuel processing has led to some highly innovative new directions in fuel processor design.

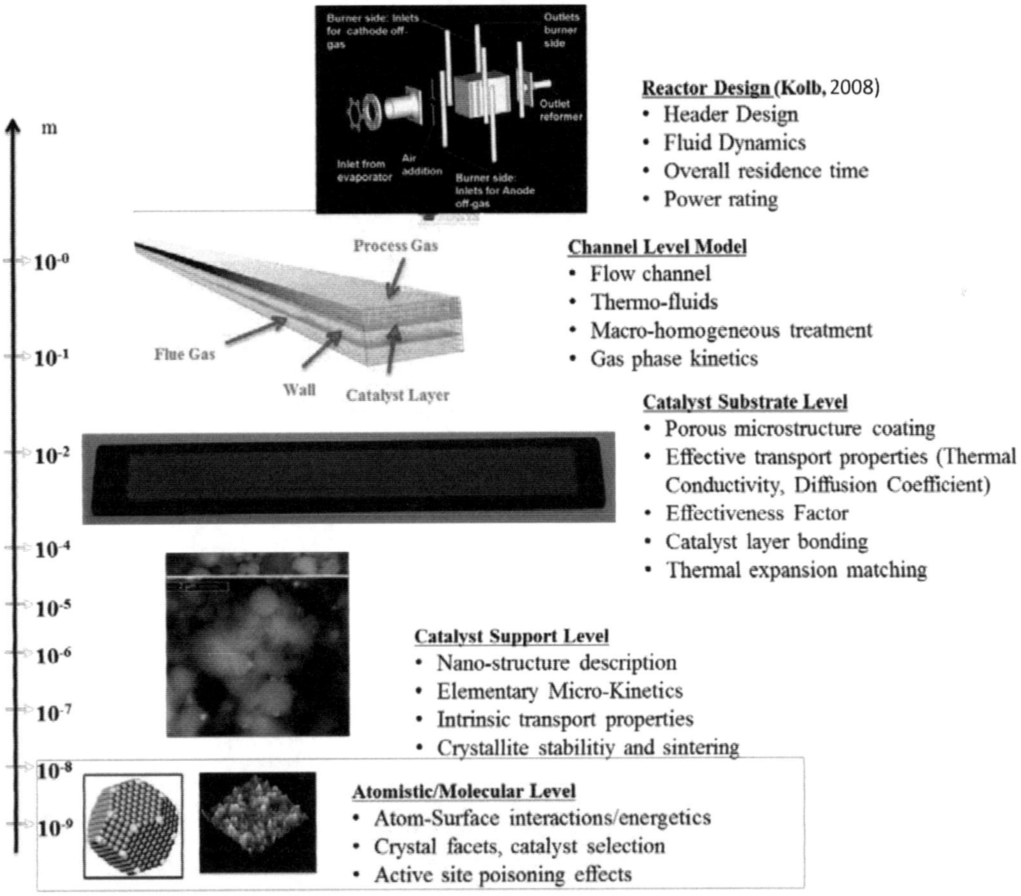

Figure 3: Understanding reactor design using a multi-scale approach where fundamental catalyst properties at the atomic level [3] are considered in the final design at the macro-level [4] offer the opportunity for significant improvements in fuel processor designs.

New directions in fuel processor design and the development of promising catalysts that are able to endure the challenges of high temperature reforming of heavier hydrocarbons has opened the door to many new opportunities for fuel cell product development.

References

[1] Selas-Linde website, Accessed 24 Sept 2012
http://www.selas-linde.com/range_of_services/modular_constructions.php
[2] Lemons, R. A., J Power Sources, 29 (1990) 251-264. Proc of First Grove Fuel Cell Symposium
[3] Haynes, D. J., D. A. Berry, D. Shekhawat and J. J. Spivey (2008). "Catalytic partial oxidation of n-tetradecane using pyrochlores: Effect of Rh and Sr substitution." Catalysis Today 136(3-4): 206-213.
[4] Kolb, G., T. Baier, J. Schürer, D. Tiemann, A. Ziogas, S. Specchia, C. Galletti, G. Germani and Y. Schuurman (2008). Chemical Engineering Journal 138(1-3): 474-489

(343b) Desulfurization of JP-8 Fuel Using Ag-Loaded Silica Supports Under Ambient Conditions

Dat T. Tran, Electrochemistry - SEDD, U.S. Army Research Laboratory, Adelphi, MD, Jessica St. John, Chemistry and Biochemistry, University of California, Santa Cruz, Santa Cruz, CA, Deryn Chu, Electrochemistry - SEDD, U.S. Army Research Laboratory, Adelphi and Scott R.J. Oliver, Chemistry and Biochemistry, University of California, Santa Cruz, Santa Cruz

2012 AIChE Annual Meeting

October 28 – November 2, 2012
Pittsburgh Convention Center

Pittsburgh, PA

Submit to **Symposium: Fuel Processing for Hydrogen Production**, as a **Talk**.

Title: Desulfurization of JP-8 fuel using Ag-loaded silica supports under ambient conditions

Dat T. Tran,[1] Jessica St. John,[2] Deryn Chu[1] and Scott R. J. Oliver[2]

[1]U.S. Army Research Laboratory, RDRL-SED-C, 2800 Powder Mill Road, Adelphi, Maryland 20783; dat.t.tran4.civ@mail.mil

[2]Department of Chemistry and Biochemistry, University of California, Santa Cruz, 1156 High Street, Santa Cruz, CA 95064; soliver@ucsc.edu

The U.S. Army demand for mobile, portable and stationary fuel cells based on hydrogen has increased rapidly. To meet these power generation demands, the U.S. Army is interested in using JP-8 fuel as a hydrogen carrier. The reformation of JP-8 to hydrogen offers an energy dense feedstock via on-demand fuel processing. Unfortunately, organic sulfur compounds intrinsic to JP-8 pose a significant challenge because they can poison the reformer and fuel cell catalysts.

In this work, mesoporous silica and aluminosilicate adsorbents loaded with silver (and other metal) ions were synthesized, characterized and examined for the adsorptive desulfurization of JP-8 fuel. The adsorbents were evaluated with JP-8 fuel containing ~700 ppmw sulfur under ambient conditions. This talk will describe the preparation and characterization of these adsorbents and their sulfur adsorption characteristics for the goal of fuel desulfurization.

(343c) Desulfurization of Commercial Fuels Using Mixed Oxide Supported Silver Adsorbents At Room Temperature for Fuel Cell Applications

A. H. M. Shahadat Hussain[1], Hongyun Yang[2] and Bruce J. Tatarchuk[1], (1)Chemical Engineering, Auburn University, Auburn, AL, (2)IntraMicron Inc., Auburn, AL

Sulfur removal from petroleum fuels is a major challenge globally due to ever growing demand of sulfur free feedstock for fuel cell applications. Adsorption of organosulfur compounds from hydrocarbon fuels at ambient conditions is a promising solution to the high energy and processing cost requirements for ultra clean fuel production. In this work, the performances of Ag adsorbents supported on titania dispersed alumina (TiO_2-Al_2O_3) for desulfurizing JP5, JP8 jet fuels and off-road diesel have been discussed. Sulfur adsorption capacities were obtained through observation of breakthrough of commercial fuels. Model fuels containing thiol and thiophene derivatives and aromatic compounds were used in breakthrough experiments to study the selectivity of the adsorbent toward sulfur species. N_2 physisorption, TPR, XRD, and *in situ* FTIR were employed to characterize the adsorbent. In the breakthrough experiments, 12% by weight Ag supported on TiO_2-Al_2O_3 adsorbent had respected saturation capacities of 11.83, 8.01, and 10.9 mg S/g sorbent for desulfurizing JP5, JP8, and off-road diesel (initial sulfur contents of 1172, 630, and 452 ppmw, respectively). Mixed oxide TiO_2-Al_2O_3 support provided seats for more dispersed and oxidized silver as observed from TPR and XRD. Breakthrough of model fuels revealed that the aromatic rings are essential for adsorption onto silver adsorbent. The sulfur capacity of the adsorbent was reduced in presence of other aromatic compounds. Infrared spectra of the samples treated with sulfur and aromatic compounds detected both physisorbed and chemisorbed sulfur species on Ag/TiO_2-Al_2O_3 adsorbent.

(343d) Multi-Physics Modeling of Auto-Thermal Diesel Surrogate Reforming

Rajesh D. Parmar[1], Amrit Jalan[2], Dushyant Shekhawat[3], William H. Green Jr.[2], Brant. A. Peppley[1] and Kunal Karan[1], (1)Queen's - RMC Fuel Cell Research Centre, Queen's University, Kingston, ON, Canada, (2)Department of Chemical Engineering, Massachusetts Institute of Technology, Cambridge, MA, (3)National Energy Technology Laboratory, Morgantown, WV

Abundant supply and existing infrastructure makes diesel a favorable candidate for portable energy supply. Diesel generators are widely used in locations where there is no reliable electric grid, but they are noisy, have emission issues, and often have poor fuel efficiency in practical operations. Because of these drawbacks, there is interest in developing Solid Oxide Fuel Cell (SOFC) systems which use diesel fuel, but which deliver electricity quietly and with lower emissions and higher efficiency.

The current study is part of a larger effort aimed at developing 1-5 kW diesel fed solid oxide fuel cells at SOFC-Canada and is mainly focused on understanding and deconvoluting the mechanism of diesel surrogate auto-thermal reforming in an experimental packed bed reactor. Gas phase kinetics play an important role during auto-thermal reforming and this study employs a detailed kinetic model developed using the automated Reaction Mechanism Generator (RMG) software. The generated model has about 9500 reactions and 450 species incorporating updated parameters from experiment and theory. The model has been validated against ignition delay data at different equivalence ratios and was found to perform reasonably well. The model has also been validated against the packed bed reactor steady state concentration data at different operating conditions. Coupling the fluid dynamics and heat transfer effects defined above with the large number of reactions and species was found to be very difficult using currently available commercial software such as Fluent and COMSOL. Hence an iterative approach was used in which simplified packed bed plug flow reactor model with heat transfer was solved using a finite element solver while the kinetics equations were solved using the CHEMKIN plug flow solver. The generated model shows the importance of entrance region effects for auto-thermal reformer design. Gas phase oxidation/pyrolysis consumes a large part of the hydrocarbon leading to lower molecular weight products that reach the catalyst surface and participate primarily in steam reforming reactions dominant on the surface of the catalyst.

(343e) Heat-Integrated Steam Reformer for Hydrogen Production

Saurabh A. Vilekar, Christian Junaedi, Kyle Hawley, Dennis Walsh and Subir Roychoudhury, Precision Combustion, Inc. (PCI), North Haven, CT

Precision Combustion, Inc (PCI) has been developing a novel heat-integrated steam reforming reactor for efficient syngas (i.e., H_2 and CO) generation based on its patented Microlith® technology, providing higher heat flux compared to other state-of-the-art reactors. The Microlith based reactor design results in a remarkably compact, lightweight, and efficient steam reforming reactor, which is otherwise widely known to be limited by the heat transfer resistance from the exothermic (heat source) to the endothermic reforming side. The burner component comprises of catalytic oxidizer instead of flame-stabilized combustion, resulting in improved temperature control and uniformity. In this paper, results from testing the reformer with a distillate fuel (containing 2-5 ppm_w sulfur) and natural gas will be presented, where near equilibrium product distribution was observed. The effect of steam-to-carbon ratio, operating pressure (i.e., up to 10 atm), and thermal inputs on the catalyst performance, fuel conversion and H_2 production will be discussed. The sulfur tolerance studies of the steam reforming catalyst and the burner catalyst will also be presented. Catalytic sulfur tolerance allows effective utilization of poor quality or waste fuels, thus significantly improving the overall process efficiency. Results from hundreds of hours of testing of the steam reformer unit showed minimal catalyst degradation, thus providing a measure of the reactor performance under realistic conditions and demonstrating the durability of the reformer. Finally, downstream hydrogen separation approaches are being examined and results will be reported. We will also discuss modeling analysis for system integration for high purity hydrogen generation.

(343f) Diesel Reforming Over Pyrochlore-Based Monolith

Daniel J. Haynes[1], David Berry[2], Dushyant Shekhawat[1], Mark W. Smith[3], Matthew M. Seabaugh[4], Richard Long[4] and James J. Spivey[5], (1)National Energy Technology Laboratory, Morgantown, WV, (2)NETL, Department of Energy, Morgantown, WV, (3)URS Corporation, Washington Division, Morgantown, WV, (4)NexTech Materials, Lewis Center, OH, (5)Chemical Engineering, Louisiana State University, Baton Rouge, LA

Diesel Reforming over Pyrochlore-Based Monolith

Daniel J. Haynes[1], David A. Berry[1], Dushyant Shekhawat[1], Mark W. Smith[2], Mathew Seabaugh[3], Rich Long[3], James J. Spivey [4]

[1] National Energy Technology Laboratory (NETL), Morgantown, WV 26505

[2] URS Washington Division, Morgantown, WV 26505

[3] NexTech Materials Ltd., Lewis Center, OH 43035

[4] Louisiana State University, Baton Rouge, LA 70803

NETL has been conducting research in developing a highly active and selective catalyst for the reforming of heavier hydrocarbon feed stocks (i.e. diesel) into a hydrogen-rich synthesis gas. This is a necessary enabling technology for the commercialization of fuel cells among other applications. Previously, NETL demonstrated 1000 hrs of continuous diesel reforming using a pyrochlore-based catalyst in powder form. However it is essential to develop a structured form, which can handle high gaseous throughput with minimal pressure drop. Key issues in the development of these structured forms include: homogeneity of coating layers, adhesion of catalyst layers (coating stability), thermal stability, and ensuring accessibility of the active sites on the catalytic material. In this study, a square channel alumina based monolith structure (400 cpi) is first coated with an oxygen-conducting support onto which an active pyrochlore phase was then deposited. These monoliths were tested for the partial oxidation (POX) and oxidative steam reforming (OSR) of pump diesel fuel. While no deactivation was observed over the duration of the tests (24h), the catalyst coating layer was not optimized and therefore the lack of homogeneity resulted in insufficient use of active phase. Imaging of the catalyst shows that there is less than optimum uniformity between the ZDC and the Pyrochlore, and current efforts are directed toward improving the contact between these two components of the catalyst.

(343g) Carbon Deposition Over Supported Rh-Ni Catalysts in Steam Reforming of Liquid Hydrocarbon Fuel for H2 Production: Influence of Support, Temperature, and Sulfur

Chao Xie[1], Yongsheng Chen[2] and Chunshan Song[1], (1)Energy and Mineral Engineering, Pennsylvania State University, University Park, PA, (2)Energy and Mineral Engineering, Penn State, State College, PA

Carbon Deposition over Supported Rh-Ni Catalysts in Steam Reforming of Liquid Hydrocarbon Fuel for H_2 production: Influence of Support, Temperature, and Sulfur

Chao Xie[a], Yongsheng Chen[a,b], Chunshan Song[a,b*]

[a] Clean Fuels and Catalysis Program, EMS Energy Institute, The Pennsylvania State University, 209 Academic Projects Building, University Park, PA, 16802, USA..

[b] Department of Energy and Mineral Engineering, The Pennsylvania State University, University Park, PA, 16802, USA.

*Corresponding author: E-mail: csong@psu.edu; Tel: 814-863-4466; Fax: 814-865-3573

This contribution aims at elucidating the effects of support (Al_2O_3, CeO_2-Al_2O_3, and CeO_2-SiO_2), reaction temperature (550 and 800 °C) and the presence of sulfur (350 ppm by weight) on the carbon deposition over bimetallic Rh-Ni catalysts in steam reforming of liquid hydrocarbon fuel for H_2 production. The amount, morphology, and chemical nature of carbon deposits on used catalysts were studied by temperature-programmed oxidation (TPO), scanning electron microscopy (SEM), and X-ray absorption near edge structure (XANES) spectroscopy. It was revealed that (I) CeO_2 modification facilitates carbon removal due to its superior redox property, and (II) Al_2O_3 is much more effective than SiO_2 in suppressing the growth of filamental carbon. As revealed by X-ray Diffraction (XRD) and temperature-programmed reduction (TPR), the Ni particles in Rh-Ni/CeO_2-Al_2O_3 were remarkably smaller than those in Rh-Ni/CeO_2-SiO_2 very likely because of the stronger interaction of Ni-Al_2O_3 as compared with that of Ni-SiO_2. Therefore, the growth of filamental carbon on Rh-Ni/CeO_2-Al_2O_3 was greatly hindered as carbon deposition becomes less favorable on smaller Ni particles. Comparative study of the used Rh-Ni/CeO_2-Al_2O_3 catalysts after the reactions without and with sulfur at 550 and 800 °C indicated that both reaction temperature and the presence of sulfur dramatically affected the catalyst behavior of carbon deposition: (I) in the absence of sulfur, the formation of filamental carbon was significant at 550 °C, which was prohibited by increasing the temperature to 800 °C, and (II) for the sulfur-containing reaction, sulfur was able to suppress the growth of filamental carbon at 550 °C, while catalyst deactivation caused by sulfur poisoning gave rise to abundant carbon deposits at 800 °C as a result of severe hydrocarbon pyrolysis at such a high-temperature condition.

Fuel Processing with Diesel Fuel and Kerosene for APU Applications

ID# 265665

Joachim Pasel[1], Jan Meißner[1], Remzi Can Samsun[1], Ralf Peters[1], Detlef Stolten[1,2]
[1]Forschungszentrum Jülich GmbH, Institute of Energy and Climate Research, IEK-3: Electrochemical Process Engineering, 52425 Jülich, Germany
[2]Chair for Fuel Cells, RWTH Aachen University, Germany

Introduction

The fuel processing unit of a fuel cell system with high temperature polymer electrolyte fuel cell (HT-PEFC) consists of three main components: the autothermal reformer (ATR), the water-gas-shift reactor (WGS) and the reactor for catalytic combustion (CAB). In combination with necessary balance-of-plant components such as pumps, blowers, heat exchangers and sensors as well as an HT-PEFC, a complete fuel cell system emerges, which can be used as an auxiliary power unit (APU) for onboard power supply in aircraft or trucks. The main benefits of fuel cell systems in comparison to conventional technologies are fuel savings and emission reduction. The following chemical reactions take place in the three main fuel processing reactors:

Autothermal reformer:

$C_nH_m + n\,H_2O$	$\longleftrightarrow$	$n\,CO + (m/2 + n)\,H_2$	endothermic
$C_nH_m + n/2\,O_2$	$\longleftrightarrow$	$n\,CO + m\,H_2$	exothermic
$CO + H_2O$	$\longleftrightarrow$	$CO_2 + H_2$	exothermic
$CO + 3\,H_2$	$\longleftrightarrow$	$CH_4 + H_2O$	exothermic

Water-gas-shift reactor:

$CO + H_2O$	$\longleftrightarrow$	$CO_2 + H_2$	exothermic

Catalytic burner:

$CH_4 + 2\,O_2$	$\longleftrightarrow$	$CO_2 + 2\,H_2O$	exothermic
$H_2 + 0.5\,O_2$	$\longleftrightarrow$	H_2O	exothermic
$CO + 0.5\,O_2$	$\longleftrightarrow$	CO_2	exothermic

Reactors for Fuel Processing

In this section, this extended abstract will focus on the methodology for the up-scaling of reactors for the water-gas-shift reaction. In recent years, three reactor generations were developed at Jülich (WGS 3, WGS 4 and WGS 5). Their progression was based on former investigations and on relevant data from the literature (1–3). Each reactor is equipped with a precious metal catalyst supported on ZrO_2-CeO_2 supplied by Clariant AG. Their design is characterized by two stages working adiabatically at different temperature levels. The high-temperature shift (HTS) stage covers a temperature range between 400 °C and 450 °C, the low-temperature shift (LTS) stage exhibits temperatures between 300 °C and 330 °C. Cooling between the two stages is achieved by quenching water into the gas stream.

The starting point of development was WGS 3 with a thermal power of 13 kW, a power density of 5.0 kW/l, and a specific power of 4.5 kW/kg. The value for the thermal power refers to the lower heating value of moles of hydrogen produced via autothermal reforming flowing through the water-gas-shift reactor. In the case of WGS 4, the thermal power was increased to 28 kW by linearly increasing the catalyst volume. The values for the power density and the specific power remained almost constant in comparison to those of WGS 3. WGS 5 with a thermal power of 140 kW was specifically developed for aircraft applications. In this case, the values for the power density and the specific power were higher and amounted to 8.5 kW/l and 8.8 kW/kg, respectively.

CFD simulations were performed on up-scaling WGS 4 to WGS 5 in order to design the area in which water is injected into the low-temperature shift stage. This area is called the evaporation chamber. Figure 1 (left) shows a sectional drawing of the first version of the WGS 5 evaporation chamber. In this version, water is injected into the evaporation chamber in liquid form at four equidistant points around the circumference. The reformate from the HTS stage flows into the evaporation chamber at a temperature of approx. 450 °C. Internal components for mixing the reformate flow with the evaporated water or for homogenizing the flow are not part of this design. However, the resulting temperature distribution in the evaporation chamber is unacceptable (see Figure 1, right). At the outlet of the evaporation chamber on the right-hand side, strong radial temperature differences between approx. 240 °C and 350 °C are clearly visible. If the catalyst in the low-temperature shift stage was fed a gas mixture with such a strong radial temperature gradient, a uniform reaction flow would be impossible.

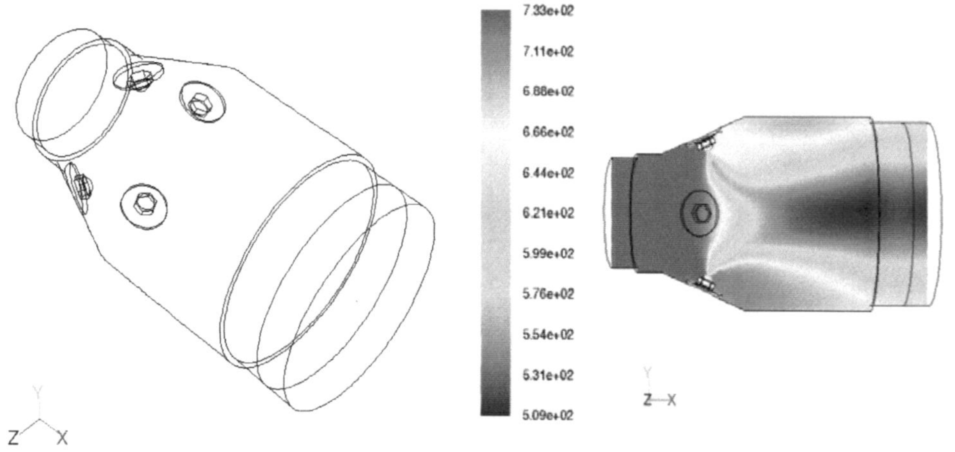

Figure 1. Sectional drawing of the first version of the evaporation chamber (left), temperature distribution in the evaporation chamber (right)

Two constructive measures were therefore implemented to prevent this temperature gradient forming. The first involved the integration of a rotating body in the flow direction above the water injection in order to ensure that the gas flow coming from the high-temperature shift stage rotates around its main flow axis. The second measure was the incorporation of a perforated plate in front of the gas flow outlet from the evaporation chamber into the low-temperature shift stage in order to homogenize the flow. Figure 2 (left) shows the second version of the evaporation chamber with the two internal components.

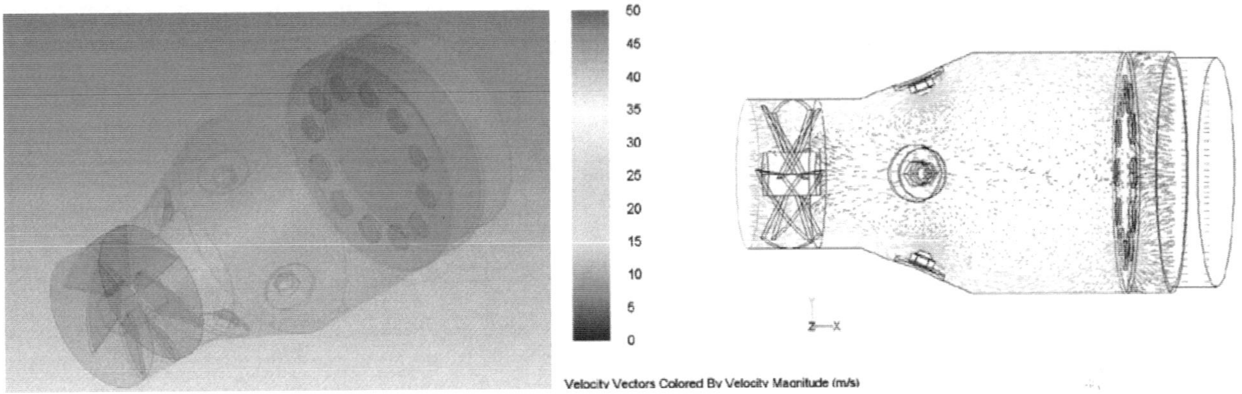

Figure 2. Sectional drawing of the second version of the evaporation chamber (left), flow profile in the evaporation chamber at 100 % load (right)

Figure 2 (right) shows the flow profile in the evaporation chamber at 100 % load produced with the aid of the components described above. The velocity vectors of the individual gaseous molecules are shown. The rotating body leads to the formation of a gas phase flow along the external wall of the reactor as a result of the centrifugal force exerted on the individual gaseous molecules. It also causes an area to develop in the center of the evaporation chamber with a backflow of some of the gaseous molecules. The perforated plate contributes to the formation of this area. The evaporation of the injected water droplets thus occurs in the counter flow: the water droplets collide with the returning gaseous molecules and evaporate. This prevents water droplets being transported to the catalyst, and a compact central evaporation zone is created.

Figure 3. Temperature profile in the evaporation chamber at 100 % load (left) and at 25 % load (right)

Figure 3 shows the resulting temperature profiles in the evaporation chamber at 100 % and 25 % load. In both cases, the gas flow had a constant temperature of approx. 300 °C at the outlet of the evaporation chamber. Under these constant conditions, a water-gas-shift reaction is possible in the low-temperature shift stage with the necessary carbon monoxide conversion.

Experimental Results and Discussion

The experimental section of this abstract deals with a number of experiments performed with the autothermal reformer ATR 8 using different kinds of diesel fuel. The fundamental design of the autothermal reformers developed at Jülich is explained in detail in [4]. The chemical and physical characteristics of the different fuels are shown in Table 1. It becomes obvious that they differ from each other in terms of their boiling ranges, their mass proportions of aromatics, and their molecular formula. For these experiments, the autothermal reformer was run at a partial load of 60 % and educt water was not evaporated or preheated before entering the internal heat exchanger of the reactor.

Table 1. Different kinds of diesel fuel applied for autothermal reforming with ATR 8

	Supplier	90 % recovered [°C]	Aromatics [wt.%]	Chemical formula
EcoPar diesel	EcoPar Sweden	338	< 0.1	$C_{15}H_{32}$
NExBTL diesel	NESTE Oil Finland	295	< 0.1	$C_{17}H_{34}$
Winter grade diesel	Shell MDS Malaysia	307	< 0.1	$C_{16}H_{34}$
Summer grade diesel	Shell MDS Malaysia	350	< 0.1	$C_{18}H_{38}$
Ultimate diesel	ARAL Germany	350 – 360	approx. 15	$C_{19}H_{38}$

Figure 4 shows the concentrations of hydrogen, carbon monoxide and methane during operation of ATR 8 with the five different diesel fuels from Table 1 at a molar O_2/C ratio of 0.47 and a molar H_2O/C ratio of 1.9. Only small differences were observed in the gas concentrations in the reformate. The hydrogen concentrations were in the range of 36.5–37.5 vol.%, those of carbon monoxide varied between 9.2 vol.% and 11.2 vol.%, while the methane contents were stable showing values of 0.3 vol.%. However, the experiment with Ultimate diesel was an exception. A remarkable decrease in the hydrogen content to 33.7 vol.% in parallel to the highest value for the carbon monoxide concentration at 11.3 vol.% were observed. This finding can be explained by the high mass fraction of aromatics contained in Ultimate diesel. For stoichiometric reasons, the conversion of aromatics during autothermal reforming delivers less moles of hydrogen than that of aliphatic hydrocarbons.

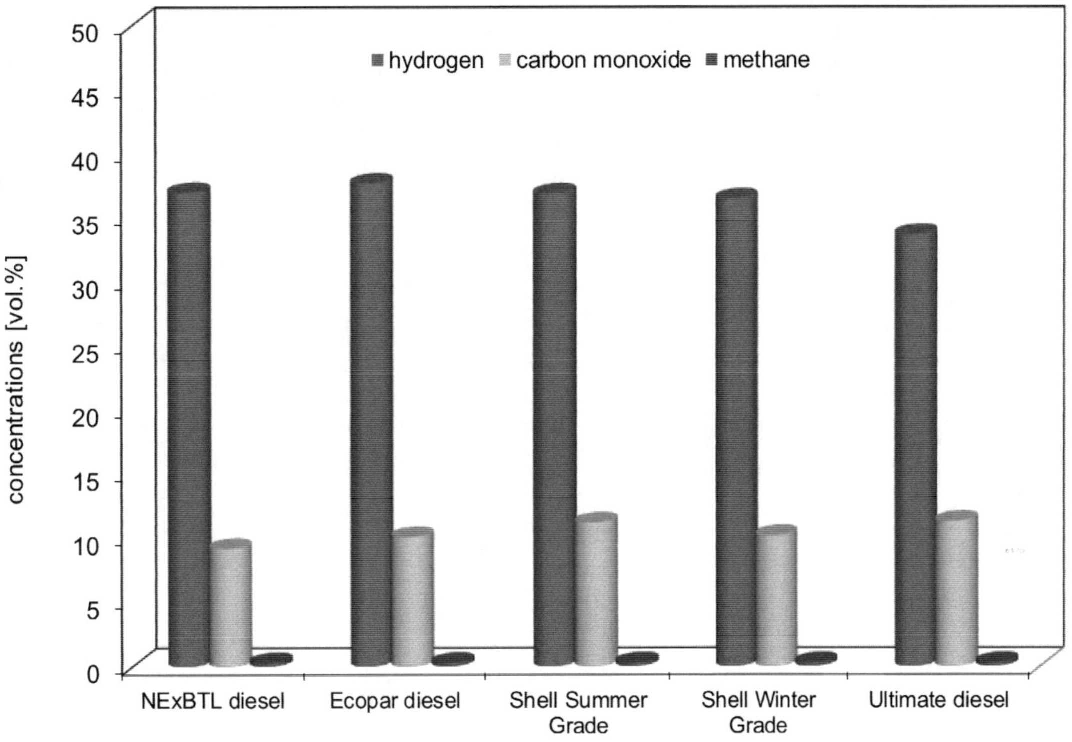

Figure 4. Concentrations of hydrogen, carbon monoxide and methane during the operation of ATR 8 with five different diesel fuels at a molar O_2/C ratio of 0.47 and a molar H_2O/C ratio of 1.9, 60 % reformer load, water injection at ambient temperature

In addition to the results shown in Figure 4 with respect to the main products obtained during operation of ATR 8 with five different diesel fuels, Figure 5 displays the respective concentrations of by-products in the gas phase and in the liquid phase at a molar O_2/C ratio of 0.47 and a molar H_2O/C ratio of 1.9. The abbreviation NMHC in Figure 5 stands for the concentration of non-methane hydrocarbons, i.e. the sum of the contents of several single by-products such as ethene, ethane, propene, propane, butenes, pentenes, hexenes and benzene. Among them, propene showed the highest values and is therefore exemplarily displayed in Figure 5. Since water is added to the reformer at large excesses (molar H_2O/C ratio = 1.9), samples of condensed, non-converted H_2O from the ATR can be collected directly behind the reactor. These samples were analyzed for the dissolved amounts of total organic carbon (TOC) and non-purgeable organic carbon (NPOC). These values are also shown in Figure 5. Acetic acid showed the highest contents among the substances in the condensed water and is also displayed. Other dissolved by-products found were acetaldehyde, propanal, acetone, butanal, 2-butanone, benzene, 2-pentanone, toluene and benzaldehyde.

Figure 5 shows very low values for the concentrations of the by-products when applying NExBTL diesel. The NMHC value was only 68 ppmv. According to Table 1, this diesel fuel is free of aromatics with the lowest temperature for 90 % recovery during fractionated distillation. When using Ecopar diesel, the concentrations for the by-products were slightly higher. The NMHC value reached

226 ppmv. A strong increase in the concentrations of all displayed by-products occurred when Shell summer grade diesel fuel was used. The NMHC value improved to 1474 ppmv, that of propene to 372 ppmv. The NPOC and TOC concentrations amounted to 134 and 170 ppmw, respectively. Acetic acid was formed at considerable amounts of 31 ppmw. Shell summer grade diesel fuel was also free of aromatics but had a remarkably higher temperature for 90 % recovery of 350 °C. Some of the single components in Shell summer grade diesel fuel demanded a higher enthalpy flow for their evaporation before entering the ATR catalyst. Therefore, the danger of liquid fuel droplets impinging on the catalyst, being submitted to pyrolysis on the catalyst, and thus damaging active sites of the catalyst is higher for Shell summer grade diesel fuel than for NExBTL diesel. Obviously, the catalyst inside ATR 8 was deactivated by the experiments with Shell summer grade diesel fuel since the concentration values for the by-products in Figure 5 when using Shell winter grade diesel fuel were much higher (NMHC = 931 ppmv) than those observed with NExBTL diesel and Ecopar diesel although their physical and chemical characteristics in Table 1 were comparable. Finally, Figure 5 shows the results obtained for Ultimate diesel, which has by far the highest mass fraction of aromatics among the fuels investigated (approx. 15 wt.%) and a comparably high temperature range for 90 % recovery during fractionated distillation of 350–360 °C. Accordingly, the concentrations for the by-products were found to be rather high. The NMHC value reached 1244 ppmv, that of propene 274 ppmv. The NPOC and TOC concentrations were 93 ppmw and 118 ppmw, respectively.

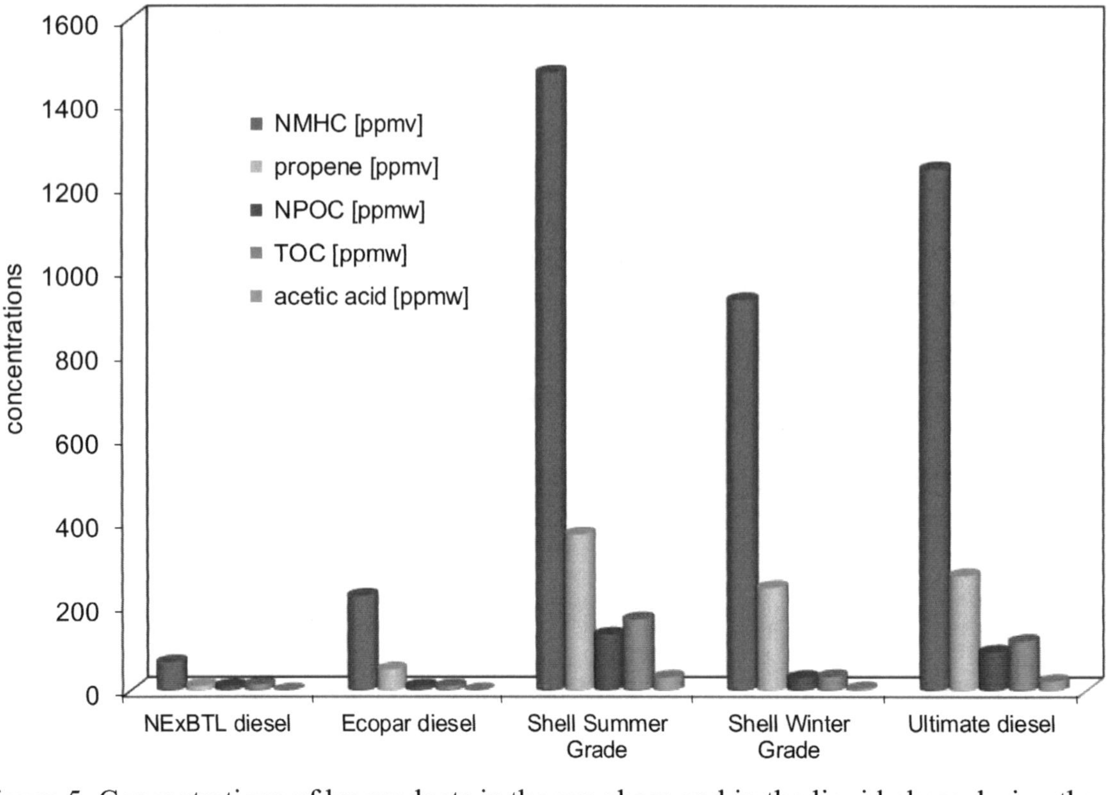

Figure 5. Concentrations of by-products in the gas phase and in the liquid phase during the operation of ATR 8 with five different diesel fuels at a molar O_2/C ratio of 0.47 and a molar H_2O/C ratio of 1.9, 60 % reformer load, water injection at ambient temperature

Outlook

More information about fuel processing with diesel fuel and kerosene for APU applications done at Jülich will be given during the respective presentation at the 2012 AIChE annual meeting and in the following paper for the special issue of Energy & Fuels.

References

1. Pasel, J.; Samsun, R.C.; Schmitt, D.; Peters, R.; Stolten, D. Journal of Power Sources 2005, 152, 189 - 195

2. Pasel, J.; Cremer, P.; Wegner, B.; Peters, R.; Stolten, D. Journal of Power Sources 2004, 126, 112 – 118

3. Koryabkina, N.A.; Phatak, A.A.; Ruettinger, W.; Farrauto, R.J.; Ribeiro, F.H. J. Catal. 2003, 217, 233

4. Pasel, J.; Meißner, J.; Pors, Z.; Samsun, R.C.; Tschauder, A.; Peters, R. Int. J. Hydrogen Energy 2007, 32, 4847 – 4858

(343i) Reactor Design for Diesel Reforming Using a Graded-Composition Monolith

Mark W. Smith[1,2,3], David A. Berry[2], Dushyant Shekhawat[2], Daniel J. Haynes[2], James J. Spivey[2,4] and John Zondlo[2,5], (1)Chemical Engineering, West Virginia University, Morgantown, WV, (2)National Energy Technology Laboratory, Morgantown, WV, (3)URS Corporation, Washington Division, Morgantown, WV, (4)Chemical Engineering, Louisiana State University, Baton Rouge, LA, (5)Chemical Engineering, West Virginia University, morgantown, WV

An NETL-developed Rh-substituted pyrochlore catalyst successfully reformed pump diesel into a H_2- and CO-rich synthesis gas for over 1000 hrs with no signs of deactivation. Even with an already low level of precious metal loading, the possibility of decreasing catalyst cost was later demonstrated using a graded catalyst bed approach. These demonstrations were performed using catalyst powders. In order to develop a commercially representative catalyst, this study examined the application of Rh-, and less expensive Ni- and Ru-substituted pyrochlore catalysts deposited onto alumina foam monoliths. The catalyst coated monoliths were placed into the reactor in several different configurations to examine the performance of a graded monolith under the catalytic partial oxidation (CPOX) of a diesel surrogate fuel. Catalytic testing was followed by carbon characterization of the individual coated foams using temperature programmed oxidation (TPO). This approach allowed for quantification of carbon formation on each catalyst in different sections of the reactor during CPOX experiments. The results of this study confirmed the trends observed in the previous powder studies and explored additional combinations of catalyst compositions in the different reaction zones of the reactor. Further it demonstrated the viability of applying the graded-composition approach to a commercially structured monolithic reactor.

(363a) Rapid Fabrication of Binder-Free Multilayered Electrodes of Polyaniline Nanofibers and Multi-Walled Carbon Nanotubes

MD Nasim Hyder[1], Yang Shao-Horn[2] and Paula T. Hammond[1], (1)Chemical Engineering, Massachusetts Institute of Technology, Cambridge, MA, (2)Mechanical Engineering, Massachusetts Institute of Technology, Cambridge, MA

Next generations electrochemical energy storage devices require high-power, high-energy density, simultaneously. To achieve this goal, it is essential to identify novel electrode materials based on nanoparticles, nanotubes or nanofibers with decreased transport length of ions/electrons and high surface area. Nanomaterials such as carbon nanotubes and conducting polymeric nanofibers /nanoparticles with their exotic properties can significantly increase power (W/kg) and energy density (Wh/kg) in addition, to their longer life cycles and charging-discharging efficiency. We assembled free-standing multilayered thick (10 to 60 microns) electrodes of conducting polyaniline (PANi) and multiwalled carbon nanotubes (MWNTs) that are robust, binder-free using rapid spray and vacuum-assisted layer by layer techniques. The electrode microstructure reveals porous nanostructure with high surface area; Preliminary results for these highly conductive electrodes (4.4 S/cm) show high pseudocapacitance of 210 F/cm^3 and are stable over 1000 cycles. We will present our findings on the optimization of the free standing asymmetric electrode morphology and electrochemical performance in aqueous and organic electrolytes.

(363b) Tunable 3-D MnO2/Graphene Composites for Next-Generation High Energy Density Supercapacitors

Lixin Wang, Chemical Engineering, Wayne State University, Detroit, MI, Da Deng, Chemical Engineering and Materials Science, Wayne State University, Detroit, MI and K. Y. Simon Ng, Department of Chemical Engineering and Material Science, Wayne State University, Detroit, MI

Supercapacitors are promising electrochemical devices for energy storage. Supercapacitors, with advantages of high power density, excellent cyclability, high efficiency and low fabrication cost, can find important applications in different areas, such as electric vehicles, industrial equipment, intermittent energy storage and electronic devices. However, current electric double-layer capacitors have low energy density, which restrict the even broader applications of supercapacitors. In the last decade, significant amount of effort has been invested on increasing the energy density of supercapacitors, specifically, the introduction of nanomaterials as electrode materials with large surface area and enhanced conductivity has achieved improvement.

In this presentation, we will present our recent effort on the design and fabrication of a high-energy 3-D composite of 2-D graphene nano-sheets and 1-D MnO2 nano-wires where graphene nano-sheets provide high surface area, high conductivity as well as excellent mechanical support for high-capacitance MnO2. MnO2 serves as spacer to prevent graphene nano sheet from restacking and also provides pseudo capacitance to extent energy performance.

The 3-D composites were fabricated by facile hydrothermal method followed by modified Hummus method. The structure and composition of the 3-D composites can be easily tuned by adjusting the reaction time, temperature, concentration of precursors. The size, morphology and crystallization of MnO2 nanowires and the mass ratio between MnO2 and graphene were optimized to achieve highly improved electrochemical performance, which demonstrated the potential applications of the 3-D composite for next-generation high-energy density supercapacitors. Three electrodes and two electrodes set up were employed to test electrochemical performance of the as prepared 3-D MnO2 and graphene composites and consistent results were obtained.

(363c) The Electrocapacitive Properties of Organometallic Polymer/Multiwalled Carbon Nanotube Hybrids

Xianwen Mao, Demetra Achilleos, Fritz Simeon, Gregory C. Rutledge and T. Alan Hatton, Chemical Engineering, Massachusetts Institute of Technology, Cambridge, MA

The Electrocapacitive Properties of Organometallic Polymer/Multiwalled Carbon Nanotube Hybrids

*Xianwen Mao, Demetra S. Achilleos, Fritz Simeon, Gregory C. Rutledge and T. Alan Hatton**

There is growing interest in adapting redox-active materials for supercapacitor applications, owing to expected synergistic effects when the capacitive charging of the electrochemical double layer can be combined with a redox (pseudocapacitive) reaction. When both charge storage mechanisms work in parallel, the overall capacitance of the composite electrode becomes the sum of the electrochemical double layer capacitance and the redox capacity. Various redox polymer electrodes have been developed for this application, but they generally suffer from limited chemical reversibility and lack long-term cycling stability in aqueous electrolyte solutions.

The redox responsiveness of hybrids comprised of organometallic polymers and multiwalled carbon nanotubes (MWCNT) makes it possible to develop supercapacitors with the desired redox capacity. We have optimized the composition of the polymer/MWCNT hybrids to achieve enhanced performance and stability as electrochemical capacitors. In this paper, we report on the morphology and structure of the hybrids as investigated by transmission electron microscopy, scanning electron microscopy and nitrogen adsorption/desorption isotherms, and study the capacitive performance, rate capability and long-term cycling stability of the hybrids using cyclic voltammetry and galvanostatic charge/discharge techniques. The charge-storage mechanism is examined based on the results of electrochemical impedance spectroscopy.

(363d) Activated Carbons Derived From Orange Peel Wastes As Electrode Materials In Electrochemical Double Layer Capacitor

Arenst Andreas Arie, Chemical Engineering, Parahyangan University, Bandung, Indonesia and Joong Kee Lee, Clean Energy and Chemical Engineering, University of Science and Technology, Daejeon, South Korea

As an electrical energy storage device, electrochemical double layer capacitor (EDLC) finds various applications in electronic products and alternative power source due to its higher energy density, fast discharge/charge time, safety and long-term operation stability. In this work, orange peel was utilized as the precursor for activated electrodes in EDLC. The surface of the carbon material was treated with the oxidative chemical agents, ZnCl2. The surface morphology, surface composition, and surface area of the as-prepared carbon materials were observed by scanning electron microscope, X-ray photoelectron spectroscopy and Brunauer–Emmett–Teller Analysis, respectively. Electrochemical characteristics of the carbon sample were studied by cyclic voltammetry, galvanostatic charge/discharge, and electrochemical impedance spectroscopy measurements in LiPF6 aqueous solution. Compared with the non-activated carbon, activated carbon showed superior capacitive performance. The activation carbon presented a high specific gravimetric capacitance of 180 F g−1. The good electrochemical performance of the activated carbon was attributed to the high surface area and larger pore volume.

(363e) Lignin Utilization in Production of Electrospun Carbon Nanofibers with Potential for Energy Storage Applications

Chuilin Lai[1], Lifeng Zhang[1], Hao Fong[1] and **Lew P. Christopher**[2], (1)Chemistry, South Dakota School of Mines & Technology, Rapid City, SD, (2)Center for Bioprocessing Research & Development, South Dakota School of Mines & Technology, Rapid City, SD

Lignin, a naturally existing biopolymer contained in lignocellulosic biomass at 20-30% on average, is a high-volume byproduct of wood pulping which in the U.S. alone exceeds 15 million tons per year. Nonetheless, lignin has been utilized mainly as a low-grade fuel for production of heat and power to date. There is however a growing demand to identify applications with higher economic value for lignin. Although production of carbon fibers from lignin has been reported in literature, the low mechanical properties obtained so far have prevented their commercial use in reinforcement and/or composites applications. Nano-fibrous felts consisting of porous carbon nanofibers derived from low-cost lignin, however, have potential applications that rely on large specific surface area such as catalyst support, supercapacitor electrode, and hydrogen storage materials. In this research, innovative carbonaceous nano-fibrous felts with high specific surface area were developed by electrospinning of aqueous mixtures of alkali lignin and poly(vinyl alcohol) with varied mass ratios followed by heat treatments for stabilization and carbonization. The developed carbonaceous nano-fibrous felts had up to 19.4-fold larger pore volume and 7-fold greater specific surface area than the PVA-based control prepared without use of lignin, and are therefore expected to find important commercial applications.

(363f) In-Situ Characterization of Transition Metal Nitrides for Supercapacitor Electrodes

Priyanka Pande[1], Alice E. S. Sleightholme[1], Paul G Rasmussen[1], Aniruddha Deb[2], James Penner-Hahn[2] and Levi T. Thompson[1], (1)Department of Chemical Engineering, University of Michigan, Ann Arbor, MI, (2)Department of Chemistry, University of Michigan, Ann Arbor, MI

***In-situ* Characterization of Transition Metal Nitrides for Supercapacitor Electrodes**

Priyanka Pande[a], Alice E. S. Sleightholme[a], Paul Rasmussen[ac], Aniruddha Deb[b], James Penner-Hahn[b],

Levi T Thompson*[ac]

[a] Department of Chemical Engineering

[b] Department of Chemistry

[c] Hydrogen Energy Technology Laboratory

University of Michigan, Ann Arbor, MI- 48109-2100.

E-mail: ltt@umich.edu

Early transition metal nitrides are promising candidates for use in supercapacitors due to their high electronic conductivities, surface areas (up to 200 m^2/g) and electrochemical stabilities [1,2]. Of these, V and Mo nitrides have been demonstrated to possess the highest capacitances. For example, VN with 1340 Fg^{-1} in aqueous KOH [3] and γ-Mo_2N with 380 Fg^{-1} in aqueous H_2SO_4 [4] have been reported. Further development of these materials would benefit from a better understanding of their charge storage mechanisms. In this paper we identify active species on the electrode-surface during charge-storage, present results from *in-situ* x-ray absorption spectroscopy (XAS), and suggest mechanisms that reconcile these results. Additionally we will present results from impedance spectroscopy and charge-discharge, indicating the device level performance for these materials.

The nanostructured V and Mo nitrides were synthesized via temperature-programmed-reaction of their oxide precursors with anhydrous NH_3 followed by passivation in 1% O_2/He at room temperature to form a oxygen-rich passivation layer preventing bulk oxidation on exposure to air [1]. Physical characterization was performed using BET surface area analysis and X-ray diffraction. Our previous results using electrolyte ion-isolation techniques suggested that OH^- and H^+ were the contributing species in charge storage for VN and γ-Mo_2N [4]. Chronopotentiometry at varying concentrations of the electrolyte was used to measure the open circuit potential (OCP). The OCP values were then used to establish the relationship between the charge-transferred and the active species. *In-situ* XAS experiments were carried out in the cell shown in Figure 1. The stability range was determined using cyclic voltammetry for VN in 0.1M KOH and γ-Mo_2N in 0.1M H_2SO_4 inside the XAS cell (*vs* a Pt wire reference electrode). The stable potential range was divided into potential steps and chronoamperometry was carried out while x-ray absorption spectra were collected at these voltages from highest to lowest potential and then in reverse.

The chronopotentiometry results for VN and γ-Mo_2N suggested that 0.5 and 1.7 electrons are transferred per OH^- and H^+ reacted, respectively. The *in-situ* x-ray absorption near-edge spectra (XANES) for γ-Mo_2N are shown in Figure 2. Based on the Mo-edge shift collected at the various potentials, we estimated the change in oxidation state for Mo at these voltages (Figure 3). These results indicated the following changes in the Mo oxidation state during charge-storage: $Mo^{3.6+}$ $Mo^{3.2+}$ $Mo^{2.8+}$ $Mo^{2.4+}$. The results for forward and reverse scans revealed minimal change in Mo oxidation state at a given potential, indicating the reversibility and cyclability of the material. *In-situ* X-ray absorption fine structure (XAFS) results will also be presented to show the changes in the coordination sphere. Parallel experiments with VN required recording spectra in fluorescence mode instead of the transmission mode as was used for γ-Mo_2N. The fluorescence mode is more difficult to analyze, but our preliminary interpretations indicate reversible redox reactions for VN.

Reference:

[1] Cladridge J B, York A P E, Brungs A J, Green Malcolm L H, Chem. Mater.12 (2000) 132.

[2] Wixom M R, Tarnowski D J, Parker J M, Lee J Q, Chen P -L, Song I, Thompson L T, Mat. Res. Soc. Symp. Proc. 496 (1998) 643.

[3] Choi D, Kumta P N, Electrochem. Solid-State Lett. 8 8 (2005) A418.

[4] Pande P, Rasmussen P G, Thompson L T, J. Power Sources. 207 (2012) 212.

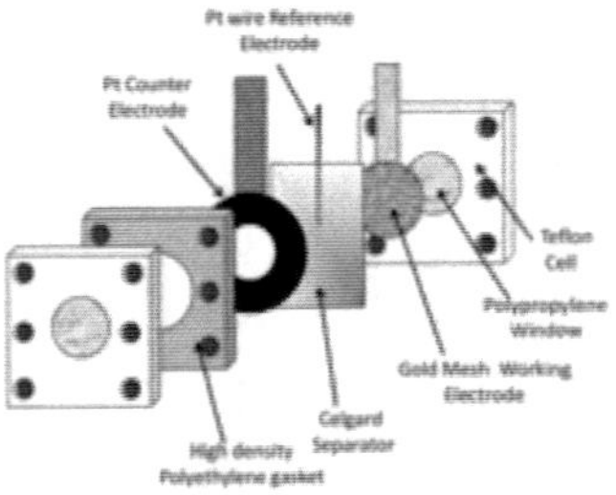

Figure 1: Schematic of *in-situ* x-ray absorption spectroscopy cell.

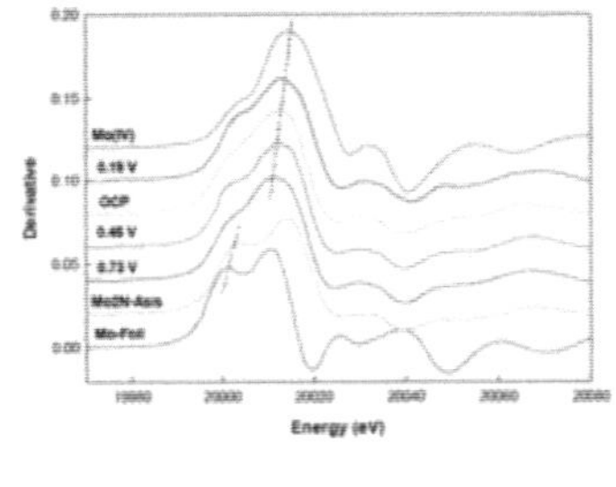

Figure 2: *In-situ* XANES spectra of γ-Mo_2N in 0.1M H_2SO_4 at various potential steps in transmission mode. Mo_2N As-is is the dry electrode. Mo-foil and Mo (IV) were used as model compounds.

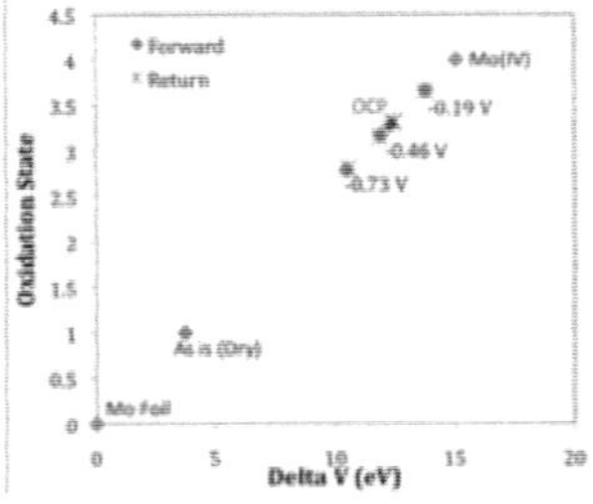

Figure 3: Changes in oxidation state of γ-Mo_2N in 0.1 M H_2SO_4 at various potentials during forward and reverse scans.

(363g) Nanostructured Vanadium Nitride for Supercapacitor Applications

Prashanth Jampani Hanumantha[1], Moni Kanchan Datta[2], Karan Kadakia[1], Dae Ho Hong[3], Michael C. Tam[1], Ayyakkannu Manivannan[4] and Prashant Kumta[1,2,5], (1)Chemical and Petroleum Engineering, University of Pittsburgh, Pittsburgh, PA, (2)Bioengineering, University of Pittsburgh, Pittsburgh, PA, (3)BioEngineering, University of Pittsburgh, Pittsburgh, PA, (4)National Energy Technology Laboratory, (5)Mechanical Engineering and Materials Science, University of Pittsburgh, Pittsburgh, PA

There is a dire need to provide a sustainable solution to the growing global demand for energy. It is this incessant demand that is the primary driver fueling various new current and emerging technologies including lithium ion batteries, supercapacitors and fuel cells. Supercapacitors hold a unique place among these energy storage devices on account of their high power density and long cyclability. Supercapacitors are electrochemical charge storage devices capable of delivering large bursts of charge over a very short period of time. It is due to this unique capability that supercapacitors occupy a very important role in the growing alternative energy field. High surface area carbons, hydrated ruthenium oxide and $MnO_2$are among the primary materials being considered for supercapacitor applications. Our group has previously identified vanadium nitride (VN) as a promising electrode material for supercapacitors on account of its high energy density with capacitances as high as 1300 F/g.

In this work, we present a facile one-step direct mechanical milling approach for the synthesis of nanostructured vanadium nitride (VN) particles. The electrochemical mechanism and material properties influencing the supercapacitor properties of VN have been explored. The surface properties of VN have also been studied using X-ray photo-electron spectroscopy (XPS) and correlated to the electrochemical characteristics. Stable capacitances of upto 100F/g are reported here for VN prepared using High Energy Mechanical Milling (HEMM). Degradation of the capacitance upon cycling is studied and possible mechanisms for the same have been proposed. Results of these studies will be presented and discussed.

(438a) Design Principles of Telluride-Based Nanowire Heterostructures for Thermoelectric Applications

Yue Wu, School of Chemical Engineering, Purdue University, West Lafayette, IN

Thermoelectric (TE) devices, which can perform a direct conversion between thermal and electrical energy, have attracted great attention due to their promising potential in improving the energy efficiency and in solid-state cooling. However, the low efficiency of the TE materials prohibits their wide applications. Certain TE materials, such as Bi_2Te_3/Sb_2Te_3 superlattice film (*ZT*~2.4) and $AgPb_mSbTe_{2+m}$ bulk crystals (*ZT*~2.2), although possessing high performance due to the improved phonon scattering at nanoscale interfaces and grain boundaries, require very complicated material composition or extremely expensive/time-consuming manufacture process such as molecular beam epitaxy. Theoretical predictions and initial experimental results have suggested that one-dimensional (1D) nanostructures, especially the nanowire heterostructures, which take the advantages of both quantum confinement to enhance the power factor and phonon scattering at nanowire surface and compositional interfaces to lower thermal conductivity, could offer a much higher *ZT* value. Meanwhile, the syntheses of various 1D nanowire heterostructures have been demonstrated through the chemical vapor deposition process based on vapor-liquid-solid (VLS) growth mechanism as well as the pulsed electrodeposition, but it is still a great challenge to obtain high-quality thermoelectric nanowire heterostuctures at a simple yet scalable way. Here we present a design principle to develop new categories of telluride-based thermoelectric nanowire heterostructures through rational solution-phase reactions. The synthesis yields "barbell" and "dumbbell" compositional-modulated nanowire heterostructures with narrow diameter and length distribution. Initial characterizations of the hot-pressed nanostructured bulk pellets of these heterostructure show a largely enhanced Seebeck coefficient and greatly reduced thermal conductivity, which lead to an improved thermoelectric figure of merit. This approach opens up new platforms to investigate the phonon scattering and energy filtering.

(438b) Excess Thermopower in Carbon Nanotubes Using Thermopower Waves

Joel T. Abrahamson, Bernat Sempere, Michael P. Walsh, Jared M. Forman, Sayalee G. Mahajan and Michael S. Strano, Department of Chemical Engineering, Massachusetts Institute of Technology, Cambridge, MA

The nonlinear coupling between an exothermic chemical reaction and a nanowire or nanotube with large axial heat conduction results in a self-propagating thermal wave guided along the nano-conduit. The reaction wave induces a concomitant thermopower wave of high power density, resulting in an electrical current in the same direction. Thermopower waves also tend to produce monopolar voltage pulses, although conventional thermoelectric theory clearly predicts bipolar voltage, and they propagate in materials with both high electrical and thermal conductivity. Temperature and voltage measurements of thermopower waves on single-walled carbon nanotubes (SWNTs) show that they can generate voltage and power in excess of the predictions of the Seebeck effect for the temperature gradient, as much as four times greater power. We hypothesize that the excess thermopower stems from a chemical potential gradient across the SWNTs with a new model that describes the electrical properties of the waves. The fuel (prototypically picramide) and ambient oxygen adsorb and dope the SWNTs ahead of the wave, and desorb and react in the hot zone behind the wave front. Raman spectroscopy confirms that picramide dopes SWNTs. Furthermore, the excess thermopower depends on the mass of fuel added (relative to SWNT mass), and the chemical potential difference matches the magnitude of the excess thermopower. This model relies on SWNT property data measured in the literature rather than fit parameters. Thermopower waves thus provide an exciting new class of highly scalable, stable energy storage devices, and offer the possibility of greatly exceeding the power limitations of conventional thermoelectric devices.

(438c) First Principles Model for Predicting the Thermoelectric Properties of Complex Antimonides and Oxides

Maria Stoica, Alireza Faghaninia, Zhou Yu and Cynthia S. Lo, Energy, Environmental and Chemical Engineering, Washington University in St. Louis, St. Louis, MO

Nanostructured inorganic materials have in the past ten years seen a rise in interest for thermoelectric applications, due to the enhanced performance exhibited at low dimensions. In our work, we study the effect of dimensionality on the thermoelectric properties of Co- and Mo- based antimonides and doped oxides. Specifically, we calculate the Seebeck coefficient, "dark" electrical conductivity, and thermal conductivity of these materials. We first perform Molecular Dynamics (MD) simulations and Density Functional Theory (DFT) calculations, where we perform Γ-point calculations as well as integrating over the first Brillouin zone. We then perform post-processing of the momentum/optical matrix elements to calculate the desired properties and predict the behavior of both known (e.g., unfilled bulk) and designed (e.g., filled bulk, nanostructured) materials. For the CoX_3 compounds (X=P, As, Sb), we look at Ag- and Au-filled skutterudites; we also compare their performance to Mo-based antimonides. For the oxides, we have performed transition metal doping of ZnO and $CuAlO_2$ compounds with oxygen vacancies, as well as direct and charge-compensated replacement of dopant atoms. We also perform computations for both bulk and 2-D nanowires for each group of materials, and demonstrate the enhancement of thermoelectric properties as a result of nanostructuring.

(438d) Characterizing Thermal Transport in Self Assembled Monolayer Junctions

Shubhaditya Majumdar[1], Scott N. Schiffres[1], Alan J. H. McGaughey[1] and Jonathan A. Malen[2], (1)Mechanical Engineering, Carnegie Mellon University, Pittsburgh, PA, (2)Department of Mechanical Engineering, Carnegie Mellon University, Pittsburgh, PA

Our investigation involves studies of the thermal transport properties of a single organic-inorganic interface – a self-assembled monolayer (SAM) junction. Molecular junctions have generated excitement as potential forerunners in organic electronics and are fundamental to the development of new hybrid materials. Characterization of their thermal properties is important in understanding how they will dissipate heat, a parasitic byproduct in their application in electronics and energy conversion. Accurate quantification of their thermal conductance, which describes this behavior, is thus essential.

Complementary experimental and computational approaches will address a prototypical Au-SAM-Au interface, where an alkane based SAM is initially considered. Molecular dynamics simulations are being implemented to model and study the effect of molecule length, choice of leads and other factors affecting bonding between the organic and inorganic components. Independent experimental measurements on transfer printed leads on SAMs are being performed using frequency domain thermoreflectance – a laser-based non-contact measurement technique. Initial measurements indicate that the Au-SAM-Au interface has a higher conductance than previously studied SAM interfaces, perhaps because prior measurements considered SAMs sandwiched between dissimilar contacts. Hence, this work is novel because the contacts are identical, and the interface conductance can be directly related to the SAM properties. This study is thus expected to provide a detailed picture on the thermal transport phenomenon across such interfaces.

(438e) Studies of Thermal Conductivity in Hybrid Organic-Inorganic Nanocrystal Superlattices

Wee-Liat Ong[1], Sara Rupich[2], Dmitri Talapin[2], Alan J. H. McGaughey[1] and Jonathan A. Malen[1], (1)Mechanical Engineering, Carnegie Mellon University, Pittsburgh, PA, (2)Chemistry, University of Chicago

Nanocrystal superlattices (NCSLs) are 3-D arrays of inorganic spheres bridged by self-assembled molecular chains. NCSL thin films hold promise for thermal management and energy conversion applications as studies have shown that their electronic and thermal transport properties can be decoupled. This unique feature allows them to be tailored according to the specific needs of the applications (e.g. high figures of merit materials for thermoelectric, high-efficiency photovoltaic materials). An additional advantage of this material comes from their solution-based fabrication technique which is readily scaled up for large volume and cheaper than conventional nanostructuring methods.

Electronic transport in NCSLs has been studied extensively. Little is known, however, about the nature of thermal transport in these complicated materials. Both experimental measurements and preliminary modelling work will be presented to elucidate the thermal transport mechanisms in NCSLs. Several factors affecting the NCSL thermal conductivity including the geometry and chemical compositions of the NCSL, will be shown.

(543a) Technology Forcing with Clean Air Act Best Available Control Technology

Mary Ellen Ternes, McAfee& Taft, Oklahoma City, OK

The Clean Air Act's Prevention of Significant Deterioration program (CAA PSD) was enacted to ensure that ambient air quality continues to meet ambient air quality standards through applicable of the Best Available Control Technology (BACT) to new major stationary sources of air pollutant emissions, or major modifications of a major stationary source. With the CAA authority being extended to greenhouse gas emissions as a result of the 2007 case Massachusetts v. EPA, and following EPA CAA greenhouse gas rulemakings, BACT as applied through PSD to major stationary sources of greenhouse gas emissions will result in driving technology and research toward methods to achieve greenhouse gas emission reductions. This presentation will discuss the existing PSD BACT program and how it has been applied recently to major stationary sources of greenhouse gas emissions.

Avoiding Backdoor Licensing and Other Tech Agreement Traps

By: Anthony P. Venturino, Esq., Partner of Novak, Druce + Quigg, LLP Washington, DC

AIChE Fall Meeting - October 31, 2012 Pittsburgh, TX

The views presented here are those of the author and not attributable to his firm or its clients.

I. Engineering is Full of Technology Agreements

Engineers may encounter agreements having technology or other forms of intellectual property in the course of business, for example organizing a startup company. Some typical agreements are listed as follows:

- Confidentiality Agreements
- Vendor Supplier Agreements
- Joint Development/Commercialization Agreements
- Exclusive Licenses / Non-exclusive Licenses
- Co-promotion Agreements
- Employee Agreements
- Settlement Agreements
- Joint Venture Agreements
- Contract R&D Agreements
- Consulting Agreements

II. Things Go Wrong

A few examples of where things can go wrong:

- Evaluation of the technology
- Selection of partners
- Conduct of the negotiations
- The agreement itself
- Implementing and servicing the agreement

Classic Licensee mistakes:

- Payments made under a trademark license when the trademark is not used
- patent royalties are still being paid even though the patents covered in the license expired or no longer cover the products
- Royalties paid at the wrong rate, i.e. one higher or lower than provided

Source: *Licensing – A Strategy for Profits*, Edward P. White (1990)

Classic Licensor mistakes:

- Failure to take action if royalty payments cease when they should not, e.g., prior to expiration of the patents.
- Failure to monitor royalty payments, e.g. sudden reductions or failure to meet minimum royalties that could trigger a change from exclusive to nonexclusive or termination of the licenses.
- Failure to stay in technical contact with the licensee, with the result of licensee not interested in paying royalties, in maintaining quality, or keeping good records.

Licensing – A Strategy for Profits, Edward P. White (1990)

III. Before you do the deal.

- Know your business goals, strengths and weaknesses
- Know the available protection for the technology
- Have a realistic idea of the value of the technology
- Be open to creative technology agreements
- Admit when you are in over your head and get help
- Do not let paranoia get in the way of a good deal
- Watch out for foreseeable legal and technical problems
- Admit when you are in over your head and get a lawyer or other licensing professional

IV. Avoiding Common Traps

A. Controlling Backdoor Licensing by University Researchers

Backdoor licensing is tech transfer traveling outside the normal or accepted route. It can be communications between a consultant and its client during the consulting. It can be an improper disclosure of information depending on contractual obligations on the discloser when the disclosure was made. For example, an employee might transfer his employer's technology to a customer in violation of his employment contract.

This can be a problem for universities. University technology transfer offices (TTOs) collect and organize invention disclosures as well as help to protect and commercialize the inventions. However, not all universities have a tech transfer office. Also, even if there is a tech transfer office, sometimes a university researcher bypasses the tech transfer office.

Taylor Aldridge and David B. Audretsch developed a database of the largest 20% of National Cancer Institute grant awardees between 1998 and 2002. Of this group 392 scientists were identified that earned patents and the authors examined how the research was commercialized. Taylor Aldridge and David B. Audretsch, "Does Policy Bias the Commercialization Route? Evidence from National Institutes of Health Funded Scientists," *Research Policy*, June 2010. The study reported 30 percent of the researchers did not assign at least one of their patentable inventions to their university employers.

Some points raised by the study are as follows:

1. "Of the 392 patenting scientists, 71.20% selected the TTO commercialization route, in that they assigned all of their patents to the university. However, 29.80% of the patenting scientists resorted to the backdoor for at least one of their patents."

2. " Scientists choosing the backdoor route for commercialization, by not assigning patents to their university to commercialize research, tend to rely on the commercialization mode of starting a new firm. By contrast, scientists who select the TTO route by assigning their patents to the university tend to rely on the commercialization mode of licensing. "

3. "The likelihood of a scientist commercializing through the backdoor may be influenced by characteristics of the TTO, the university, the region, the scientist, and the technological nature of the invention itself."

4. "[T]here is at least some evidence suggesting that the commercialization path is not neutral with respect to the significance of the potential innovation. Rather, those innovations with a potentially greater value are less likely to be commercialized through the technology transfer office."

Another type of backdoor licensing is informal technology transfer between university scientists and industry personnel. For example, a company hires a university researcher as a consultant and inventions are made on company time with company resources.

How can a University respond?

One approach is to give researchers incentives to go through the TTO. For example, universities could make patenting one criterion for promotion and tenure. Christoph Grimpe and Heide Fier, Informal University Technology Transfer: A Comparison Between the United States and Germany, Discussion Paper No. 09-033, ZEW Zentrum für Europäische Wirtschaftsforschung GmbH, Centre for Economic Research (2009); URL<ftp://ftp.zew.de/pub/zew-docs/dp/dp09033.pdf>.

Another approach is to have a patent policy and a culture encouraging disclosure. Taylor Aldridge and David B. Audretsch, indicated "There may also be systematic differences between public and private universities. Contributing to state economic development, growth, employment creation and technology transfer is often an explicit mandate at publicly funded state universities. This is rarely the case at private universities. Thus, public universities may be more sensitive to controlling, monitoring and capturing the flow of intellectual property going from university scientists to commercialization."

B. <u>Define how to handle joint invention of improvements</u>

By US statute joint inventions are jointly owned. It is better to have one party own the invention and, if necessary, grant the other party an exclusive field of use license. Thus, a clause such as the following should be included if a jointly developed invention may occur during performance of the agreement:

> In the event of a joint development my company owns the invention and controls patent prosecution and your company gets an exclusive use license in the field of _______________.

C. <u>Keep good records of what the agreement covers or when it expires</u>

The company, its legal department or a law firm should docket the agreement in a database. The docketing record should include: the type of agreement, the parties, the technology covered by the agreement, the deadlines, the expiration date, the contact person and backup contact person, and have a miscellaneous portion for anything unusual.

Preferably an institutional person, for example, a docket clerk or contracts administrator, should be responsible for tracking notices in the docketing system and then relay the information to the business people. The more automatic you can make the timely conveying of such information, the better.

D. <u>Do not let an agreement to agree accidentally become an agreement</u>

The first time you negotiate with a party, especially if the other party sees this agreement as a one shot deal, be careful not to make records during the negotiations which unexpectedly bind you. Do not accidentally create an agreement.

Summaries of the negotiations signed by both parties can lead to binding agreements even when a clause that might be important to you is missing. Let's take a hypothetical scenario where two parties bring their business people and lawyers to a meeting to negotiate a potential IP license. Mid-afternoon the attorney for the IP owner (potential licensor) had to leave to catch a plane. The potential licensor's business person (let's call him "Bill") had a later plane. The potential licensee kindly offered to take Bill to the airport later so he would not have to waste time in a boring airport. While Bill was by himself with the potential licensee's attorney and business people one of the business people suggested they draft and sign a summary of the day's discussions. They did this and then Bill caught his plane. The next day Bill's company was asked to sign an Assignment document to reflect the assignment they agreed to yesterday. Bill's response was "What assignment?" After a lawsuit the parties settled. The moral of the story is train the business people not to sign anything.

A deal can also inadvertently arise if the parties act like there is an agreement can create an agreement. This sometimes arises where the parties are in a rush to do joint development. While the deal is only a draft they start working together. Sometimes the unsigned draft becomes the *de facto* agreement.

The opposite can also happen. Sometimes you think you have an agreement but in reality you do not, or at least not all the clauses apply. Take the case of *DataTreasury v. Wells Fargo,* 522 F.3d 1368 (Fed. Cir. 2008). In 2003 WMR e-Pin licensed a patent to Wells

Fargo Services Corp. a subsidiary of Wells Fargo & Co. The license included an arbitration clause. In 2006 WMR e-Pin assigned 4 patents to DataTreasury. At least one of the 4 patents was a patent of the license. Afterwards DataTreasury sued Wells Fargo & Co. for patent infringement in court. Wells Fargo & Co. moved to dismiss or stay pending arbitration. Under 5th Circuit law, as applied by the Federal Circuit, "to be enforceable, an arbitration clause must be in writing and signed by the party invoking it. *DataTreasury* at 1373, quoting *Westmorland v. Sadoux*, 299 F.3d 462, 465 (5th Cir. 2002). Neither Wells Fargo & Co. nor DataTreasury signed the original patent license. Thus, the court held these parties were not bound by the arbitration clause.

E. <u>Remember important clauses</u>

Clause checklists are important tools when reviewing a license to see if all the necessary clauses are present. However, use the right checklist for the type of agreement and the type of IP. Do not assume the clauses in your standard patent license belong in your new trademark license and *vice versa*. Unlike patents which protect the patent holder, trademarks protect the public by assuring the public a good labeled with a particular trademark will have the characteristics they expect. For example, a PIRELLI brand tire has certain handling properties. Trademark licenses without quality safeguards are termed naked licenses and invalidate the trademark rights.

A short checklist is provided in TABLE 1:

TABLE 1
Introductory Clauses
Definitions
Grant of Rights -patents-trade secrets
Consideration/Compensation
Diligence
Product Quality Control
Advertising
3rd Party Infringement Suits
3rd Party Product Liability Suit
Licensor/ Licensee Suits
Termination
Continuing Obligations

A detailed set of clause checklists for a licensing agreement is available upon request from the author.

F. <u>Remember to protect the Trade Secrets and Know-how</u>

- Define trade secrets
- Set up a manageable way to transfer the information

G. Keep control of the technology after making the deal

Proprietary technology is a competitive advantage. If the licensor loses control of the technology he loses control of the competitive advantage. Thus, the licensor should determine the level of control he is willing to lose. Maybe the licensor thinks he can permit the licensee to sublicense because he trusts the judgment of licensee's president. What if the president leaves? Thus, the agreement preferably precludes sublicensing or assigning to competitors.

Another common way to lose control is for the licensee to be bought by the licensor's competitor. The Agreement should specify it may be assigned by ONE of the following:

1. Anyone;
2. Goes with the business;
3. Goes with the company;
4. Licensor maintains right of first refusal;
5. Licensor maintains right of first refusal which will not be unreasonably withheld;
6. No one.

H. Promote good communication after making the deal

Typically problems arise from misunderstandings, lack of thoroughness, carelessness, or failure to administer the agreement at all, i.e., no one in authority is responsible for administration (LES 1983-84 study).

The lawyer can structure the agreement to include mechanisms for communication between the licensor and licensee so little problems and misunderstandings do not become big ones. You cannot force the business people to communicate but the mechanisms provide fertile soil for communication. For example, the agreement should include a contact person and back-up contact person. The agreement can also mandate times for periodic meetings.

I. Include Clean Termination Provisions

Relationships can end badly. Thus, you need the equivalent of a pre-nuptial agreement. Include termination provisions in the agreement to give each party a reasonable way to terminate the agreement. This includes planning for unexpected events, for example, bankruptcy or purchase of the other party. This also includes providing ways to terminate or renegotiate if the economics of the deal no longer make the current agreement viable.

V. Organize A Licensing Department – Overview

Some companies only occasionally do a license or may be at the beginning of setting up a license project group. In these cases, the company might ask his or her lawyer for advice on how to organize the licensing project group.

The license project group will need the support of a top executive. Without this the license project group will generally not get the resources it needs.

Recognize the different functions to be performed. Then, divide or share these functions depending on available resources, business people, consultants or lawyers:

- Profit Center Group
  - optimizes the licensing in and licensing out to make money for the company
- Support Group
  - coordinates technical, legal and accounting aspects of tech transfer
- Administrative Group
  - Handles accounting, record keeping and day to day details of maintaining the license.

VI. Conclusion

University researchers and their Technology Transfer offices can work together regardless of whether the university researcher wants to license the invention or commercialize it herself. However, Universities and their Technology Transfer Offices should strive to have policies, procedures and culture to be a bridge to encourage partnerships with researchers and the outside industrial community.

Agreements are necessary to manage intellectual property. This applies regardless of whether the agreements involve the university researcher, the industrial entrepreneur or the Fortune 500 Company If you know your goals and avoid the traps your agreements can help turn technology into profit.

(543c) Quantifying the Microeconomic Impacts of State-Level Incentive Programs On Biorefineries

Tristan Brown[1], Rajeeva Thilakaratne[1], Guiping Hu[2] and Robert C. Brown[3], (1)Bioeconomy Institute, Iowa State University, Ames, IA, (2)Iowa State University, Ames, IA, (3)Mechanical Engineering, Iowa State University, Ames, IA

A growing number of U.S. states are implementing laws and policies that incentive the construction of biorefineries within their boundaries. These incentives range from wide-ranging Renewable Portfolio Standards (RPSs) to narrowly-drawn interest-free loans, such as the $75 million loan recently granted to catalytic pyrolysis company KiOR by the state of Georgia for the construction of a biorefinery in the state. Corporate income tax rates are another factor affecting the economic feasibility of biorefineries that vary widely across states.

We compile and assess the major state-level incentive programs currently being employed by different states to incentivize biorefinery construction. A techno-economic Chemcad model of a biorefinery employing the fast pyrolysis and hydroprocessing pathway for the production of biofuels and bioproducts is built alongside a discounted cash flow rate of return (DCFROR) spreadsheet. The effects of these incentive programs are simulated within the model and their microeconomic impacts are quantified and compared. The result ranks the major incentive programs according to both their direct costs and their net benefit (or harm) to the biorefinery.

(576a) Noble Metal-Based 1-Dimensional Nanomaterials in the Application of Direct Alcohol Fuel Cells

Liang Su, Yixin Liu, Wenzhao Jia and Yu Lei, Chemical, Materials & Biomolecular Engineering, University of Connecticut, Storrs, CT

Direct alcohol fuel cells (DAFCs) are envisaged to be a new generation of portable power sources which convert chemical energy into electrical energy with, theoretically, both economical and environmental benefit. Noble metal-based composite nanomaterials have shown great applicability as the cathode and anode catalysts in application of DAFCs. Furthermore, 1-D bimetallic nanomaterials have shown improved performance towards oxygen reduction reaction and alcohol oxidation reaction attributed to the ameliorated durability from the 1-D nanostructure and the enhanced catalytic activity from the incorporation of the second metal.

In this work, Pt-based nanomaterials with 1-D hollow nanostructure prepared from a novel, robust and versatile electrospinning–electroless plating–galvanic replacement reaction process. The overall procedure for the material preparation is also economically efficient and environmentally friendly. The as-prepared Pt nanotubes and Pt-carbon nanofibers have been characterized by SEM, TEM, XRD, and XPS. Furthermore, these nanomaterials have shown excellent catalytic activity with superior durability towards oxygen reduction reaction and ethanol oxidation reaction in the application of DAFCs.

(576b) Nanoporous, Conducting, Electrocatalytically Active, 10nm Thick Core-Shell Au@Pt Nanoparticle Bilayer Architectures for PEM Fuel Cell

Ipshita Banerjee, Chemical Engineering, Indian Institute of Science, Bangalore, India, Venugopal Santhanam, Department of Chemical Engineering, Indian Institute of Science, Bangalore, India and V Kumaran, Chemical Engineering, Indian Institute of science, Bangalore, India

The main challenge facing the development of Proton exchange membrane fuel cells (PEMFC) currently is to fabricate a catalyst layer with minimum loading of precious platinum nanoparticles, maximum availability of eletrocatalytically active area, and high durability of the catalyst layer in presence of poisonous CO species and corrosive electrochemical atmosphere. The different approaches reported in literature to prepare catalyst layers can be broadly divided into two types: 1) where the platinum nanoparticles are physisorbed on carbon substrates like carbon nanotubes, buckypaper, xerogels etc., and 2) where thin films of white gold leaf are dealloyed and electroplated with a thin platinum layer to form the electrocatalytically active catalyst membrane. In our group, we are developing a novel bottom-up approach, wherein self-assembled monolayers of alkanethiol-coated, monodispersed Au@Pt core shell nanoparticles are stacked to fabricate the catalyst layer. Herein, results on bilayer membranes of Au@Pt core shell nanoparticles will be reported. Initially, the bilayers were electrocatalytically inactive, and had a high sheet resistance of the order of GΩ, due to the presence of organic ligands. The bilayers were then treated to room temperature RF plasma in an argon environment. The results of electrical, spectroscopic and microscopic characterization show that argon plasma exposure removes the alkanethiol ligand coating and leads to coalescence of neighbouring nanoparticles, thereby rendering the bilayer electrocatalytically active and electronically conducting. After an optimized time of exposure, the individual nanoparticles in the bilayer stacked laterally coalesce together to form electrically continuous, nanoporous metallic film with sheet resistances that are eight orders of magnitude lower than the ligand-coated bilayer array. We will present the results of our studies on the electrocatalytic activity and conductivity of membranes prepared with different Pt/Au ratios, and for different durations of argon plasma exposure. The activity of these bilayer membranes towards methanol oxidation and CO oxidation will also be reported.

(576c) Ion Pair Reinforced Semi-Interpenetrating Polymer Networks for Fuel Cell Polymer Electrolyte Membrane Design

Chunliu Fang, Graduate School for Integrative Science and Engineering, National University of Singapore, Singapore, Singapore, Liang Hong, Department of Chemical and Biomolecular Engineering, National University of Singapore, Singapore, Singapore and **Jim Yang Lee**, Chemical & Biomolecular Engineering, National University of Singapore, Singapore, Singapore

Semi-interpenetrating polymer networks (SIPN) are a special class of polymer composites where one or more linear (or branched) polymers percolate molecularly into a network of other polymers. It provides a high degree of homogenization of the component polymers beneficial for the fabrication of polymer electrolyte membranes (PEMs) for fuel cell applications. For example a linear ionomer can be used as the proton source to infiltrate into a cross-linked hydrophobic polymer network which provides the mechanical properties. The interlocked polymer structure leads to good intermixing of the hydrophilic and hydrophobic components to support a distributed presence and good connectedness of the ionic clusters.

Despite these promising prospects, SIPN membranes for fuel cell applications have yet to be satisfactorily demonstrated. In many of the studies, the network polymer was cross-linked by ester linkages. The susceptibility of the ester linkages to hydrolytic degradation in the acidic fuel cell environment compromises the SIPN performance.

This study proposes a new SIPN design where ion pairs are used to reinforce the SIPN structure. The SIPN was synthesized from linear sulfonated poly(2,6-dimethyl-1,4-phenylene oxide) (SPPO), brominated PPO (BPPO), and ethylenediamine (EDA) cross-linker by a one-step thermal cross-linking procedure. Ion pairs were formed after cross-linking and strengthened the attachment of SPPO to the BPPO/EDA network. The cross-links in the membranes are chemically resistant. Measurements of membrane dimensional changes and hydrolytic stability confirmed the improvements made to the PPO membranes for fuel cell applications.

(576e) Nanoscale Membrane Design Principles for Optimizing Proton Conductivity

David J. Keffer, Chemical and Biomolecular Engineering, University of Tennessee, Knoxville, Knoxville, TN

A model for charge transport in aqueous systems is used to provide insights into design of proton exchange membranes that optimize charge transport. The model accounts for the effect of acidity, confinement and connectivity on the diffusivity and has been shown to quantitatively agree with experimental measurements of charge self-diffusivity in Nafion [Esai Selvan, M., et al. *J. Phys. Chem. B* **115** 2011 pp 3052–3061] and in cross-linked and sulfonated polycyclohexadiene. In this presentation, the origin of the theoretical model is described. The model is applied to water-filled cylindrical nanopores functionalized on their interior surface with acid groups. It is demonstrated that for cylindrical nanopores of a given radius there is an optimal surface coverage of acid groups. The optimum can be sharply peaked, indicating that non-optimal surface coverages (either too low or too high) drastically reduce the conductivity of the pore. The theoretical maximum conductivity through a cylindrical nanopore is calculated to be about 0.70 S/cm at 300 K. Comparison with available experimental results is made. Limitations of the model are also discussed.

(622a) Synthesis and Redox Pathways of CuO-Al_2O_3 Oxygen Carriers for Chemical Looping Combustion

Qasim Imtiaz[1], Agnieszka Kierzkowska[1], Marcin Broda[1], Christian Koenig[2], Olga Safonova[2], Maarten Nachtegaal[2], Tilman J. Schildhauer[2] and **Christoph R. Müller**[1], (1)Department of Mechanical and Process Engineering, ETH Zurich, Zurich, Switzerland, (2)Paul Scherrer Institute, Villigen, Switzerland

Abstract

Fossil fuels are the dominating energy carriers for the electricity generation and transportation sectors. The combustion of fossil fuels accounts for approximately one third of the total global CO_2 emissions, therefore, it is imperative to find ways to capture CO_2 from the flue gases of fossil-fuel fired power plants, oil refineries and large industrial facilities at low costs. Chemical looping combustion (CLC) is an emerging carbon dioxide capture and storage (CSS) technology that offers an efficient and economical way to produce a pure stream of CO_2 and to reduce the emission of NO_x. CLC is a two stage process comprised of a reduction and an oxidation stage. In the reduction step, lattice oxygen of the oxygen carrier is used to oxidize a (hydro-)carbonaceous fuel, say CH_4, *viz*: $CH_4 + 4CuO \rightarrow 4Cu + CO_2 + 2H_2O$. Subsequently, Cu is re-oxidized with air, $2Cu + O_2 \rightarrow 2CuO$, to close the cycle. The CLC process produces inherently a pure stream of CO_2 (after the condensation of steam). Consequently, the high costs associated with the capture of CO_2 from flue gases, mostly arising in the regeneration step of "conventional" CO_2 sorbents, are substantially reduced. Copper oxide is a very promising oxygen carrier for CLC owing to its (i) high oxygen carrying capacity, (ii) exothermic reduction reactions, (iii) low tendency for carbon deposition, and (iv) high reactivity for both the reduction and oxidation reactions. However, because of the low Tammann temperatures of CuO and Cu of only 558 °C and 405 °C, respectively, unsupported CuO is prone to agglomeration and thermal sintering. Hence, CuO is typically supported on a high melting point material, *e.g.* Al_2O_3.

The aim of this study was to investigate the synthesis steps and reaction pathways of Cu-based, Al_2O_3-supported, oxygen carriers. Here, Cu-rich (82 wt. % CuO) oxygen carriers were synthesized using a co-precipitation technique. A detailed characterization of the synthesized oxygen carriers with regards to their morphological properties, chemical composition and surface topography was carried out using powder X-ray diffraction (XRD), scanning electron microscopy (SEM), attenuated total reflection Fourier transform infra-red (ATR-FTIR) spectroscopy, Raman spectroscopy, N_2 adsorption and force gauge measurements. The redox stability of the oxygen carriers was evaluated at 800 °C in a fluidized bed reactor using methane as the fuel.[1-2] In addition, *in-situ* X-ray absorption spectroscopy (XAS) was performed at 500 °C using $H_2$and air as the reducing and oxidizing gases, respectively, to probe the reduction and oxidation pathways of the new oxygen carrier.

Raman spectroscopy revealed that Al^{3+} was precipitated as gibbsite at a pH value of 8.5 and as boehmite at all other pH values investigated. Furthermore, the Raman spectra showed that Cu^{2+} precipitated out as gerhardtite at all pH values studied. The scanning electron micrographs showed that the surface morphology of the synthesized oxygen carriers was strongly influenced by the pH value at which the precipitation was performed. Furthermore, an Al-rich oxygen carrier was formed at pH = 3.8 owing to (i) an insufficient amount of NaOH to achieve complete precipitation of Cu^{2+} and Al^{3+} ions from the solution and (ii) the smaller solubility constant of $Al(OH)_3$ ($K_{sp} = 4.6 \times 10^{-33}$)[3] compared to $Cu(OH)_2$ ($K_{sp} = 2.2 \times 10^{-20}$)[3]. On the other hand, the chemical composition of the oxygen carriers synthesized at pH values ≥ 5.5 was similar. All oxygen carriers showed a stable cyclic oxygen carrying capacity. XRD analysis revealed that CuO, present in the form of a mixed oxide ($CuAl_2O_4$) in the calcined oxygen carriers, is fully reducible. Therefore, the formation of $CuAl_2O_4$ did not reduce the oxygen carrying capacity of the synthesized oxygen carriers. With respect to reaction pathways, previous, low-temperature (200 – 300 °C) *in-situ* XAS measurements by Kim *e. al.* revealed that the reduction of commercial, pure CuO using a gas mixture containing 5 % H_2 (flow rate > 15 mL/min) proceeded *via* the direct CuO – Cu transition.[4] In contrast, the preliminary *in-situ* XAS measurements reported here, indicate that at 500 °C the reduction of Al_2O_3-stabilized CuO proceeded *via* the sequential transition, i.e. CuO – Cu_2O – Cu. Also, using *in-situ* XAS measurements the induction period for the reduction reaction of co-precipitated CuO was found to be negligible at 500 °C. Furthermore, we found that under the operating conditions studied here, the Cu_2O intermediate did not fully reduce to Cu (even after a reaction time of 15 mins.). Similar observations, however at lower operating temperatures, were reported by Kim et al. who explained this observation by the formation of an impermeable film of Cu on Cu_2O grains. With regards to the oxidation reaction, it was found that at 500 °C also the oxidation of Cu/Al_2O_3 proceeded *via* the Cu_2O intermediate. Thus, *in-situ* XAS measurements demonstrated experimentally the formation of Cu_2O in both the reduction and oxidation step at operating conditions relevant for the chemical looping process. This finding has important consequences with regards to the numerical modeling of the chemical looping process.

[1] Q. Imtiaz, A. M. Kierzkowska and C. R. Müller, *ChemSusChem* **2012**, DOI: 10.1002/cssc.201100694

[2] Q. Imtiaz, A. M. Kierzkowska, M. Broda and C. R. Müller, *Environ. Sci. Technol.* **2012**, 46, 3561.

[3] R. W. Clark and J. M. Bonicamp, *J. Chem. Ed.* **1998**, 75, 1182.

[4] J. Y. Kim, J. A. Rodriguez, J. C. Hanson, A. I. Frenkel and R. L. Lee. J. Am. Chem. Soc. **2003**, 125, 10684.

(622b) Novel Ferrite Materials for H2 Production Using Chemical Looping

Victoria J. Aston, Brian W. Evanko, Jonathan Scheffe and Alan W. Weimer, Chemical and Biological Engineering, University of Colorado at Boulder, Boulder, CO

The chemical looping hydrogen (CLH) production process is an alternative method to water-gas shift to generate pure H_2, as well as a separate stream of pure and easily-sequestered CO_2, from any synthesis gas without complicated and expensive separations equipment. Synthesis gas is used to reduce a metal oxide material at moderate to high temperatures as it leaves a gasifier (500°C to 800°C), producing a stream of CO_2 and H_2O. The resulting reduced metal is then oxidized with H_2O to produce H_2 and regenerate the metal oxide. This process requires use of a metal oxide material that achieves high H_2 and CO conversions to H_2O and CO_2 during the reduction step. Complete regeneration of the metal oxide using only H_2O as an oxidant is also desirable. In this study, the mixed metal ferrites $CoFe_2O_4$ and $NiFe_2O_4$ have been investigated as metal oxide candidates for the CLH process. The ability of these ferrites to oxidize H_2 and CO in the syngas to H_2O and CO_2 during the reduction step was compared to that of the Fe_2O_3 in a packed bed reactor and found to be nearly equivalent. Unlike Fe_2O_3, the mixed metal ferrites were regenerated with H_2O oxidation only, allowing all the H_2 and CO used during the reduction step to be recovered as H_2 during oxidation. Experimental results also show the H_2 production capacity of the reduced ferrites to be much greater than that of the Fe_2O_3 per cycle under the same conditions. Long-term cyclability tests showed the $CoFe_2O_4$ material to cycle with great consistency.

(622c) Reducible Supports for Nickel-Based Oxygen Carriers in Chemical Looping Combustion

Saurabh Bhavsar[1,2] and Götz Veser[1,2], (1)Department of Chemical and Petroleum Engineering, University of Pittsburgh, Pittsburgh, PA, (2)US DOE-National Energy Technology Laboratory, Pittsburgh, PA

'Chemical looping combustion' (CLC) is an emerging technology for clean energy-production from fossil and renewable fuels. In CLC, an oxygen carrier (typically a metal) is first oxidized with air. The hot metal oxide is then reduced in contact with a fuel in a second reactor, thus combusting the fuel. Finally, the reduced metal is transferred back to the oxidizer, closing the materials "loop". In this way, CLC produces sequestration-ready CO_2-streams without expensive air separation and hence without significant energy penalty. Combined with sequestration, CLC thus allows high-efficiency, CO_2 emissions-free combustion of fossil fuels, or combustion processes with negative CO_2-footprint from biomass-derived fuels.

To-date, various oxygen carriers have been investigated to maximize carrier and fuel conversion while making the carriers more thermally and mechanically stable to obtain sustained performance at demanding operating conditions of CLC. Potential oxygen carrier metals viz. Ni, Fe, Cu, Co and Mn have been studied extensively. But these metal oxides have been investigated only on non-reducible supports like Al_2O_3, SiO_2, TiO_2, etc. some of which are currently considered to be most viable supports for CLC.

Here, we report on comparative evaluation for performance of Ni-based oxygen carriers supported on reducible supports (CeO_2, La_2O_3) and non-reducible supports (Al_2O_3, SiO_2). Most of the published work to-date has been focused on using only non-reducible supports. But using reducible supports not only leads to understanding their effect on reactivity of supported metal in redox operation, but also allows estimation of added contribution of support for oxygen carrying capacity of carrier.

In this work, we synthesize Ni-based carriers (40 wt% Ni) by simple incipient wet impregnation technique on as synthesized and/or commercial supports. These as synthesized and post reactive tests carrier samples are characterized by TEM, XRD, BET, and EDX. Initially, the reducible supports CeO_2 and La_2O_3; along with Ni-based carriers were tested in thermogravimetric analyzer (TGA) at 800°C for their activity in CLC with H_2 as a model fuel. Although CeO_2 and La_2O_3, themselves do not contribute significantly to overall oxygen carrying capacity. More importantly, Ni-CeO_2 and Ni-La_2O_3 show complete carrier utilization in TGA, whereas Ni-SiO_2 and Ni-Al_2O_3 shows partial utilization of oxygen carrying capacity (~91% and ~81% respectively). With increase in operating temperature, contribution from reducible support shows a marginal increment and non-reducible support show improvement in carrier conversion.

Ni-CeO_2 and Ni-Al_2O_3 are then down selected based on parameters like reducibility, thermal stability and cost for further evaluation with CH_4 as fuel in TGA and fixed-bed studies at 800°C. The experiments were aimed at limiting/avoiding coking on the carriers, critical for their long term reactivity in CLC, by means of limited CH_4 exposure for reduction half cycle. In TGA studies, Ni-CeO_2 again shows complete carrier utilization and thus outperforms Ni-Al_2O_3 (~77% carrier conversion). Post use characterization of carriers by TEM, XRD, and BET does not reveal any significant sintering and spinel formation for the two carriers. Furthermore, these carriers are then evaluated in fixed-bed reactor, wherein the reduction period was limited to complete combustion of CH_4. Ni-CeO_2 not only shows a longer period of reduction half cycle indicating better carrier performance, but also >90% fuel conversion and thus is a promising candidate for efficient CO_2 capture via CLC. Thus our results show, reducible support based carriers can be a new class of cheap and efficient carriers for CLC operation.

(622d) Cu Oxidation Kinetics In a Chemical Looping Combustion Systems

Eli A. Goldstein and Reginald E. Mitchell, Mechanical Engineering, Stanford University, Stanford, CA

Knowledge of intrinsic chemistry will greatly facilitate the optimization of fuel and air reactors in Chemical Looping Combustion (CLC) power systems. In order to determine intrinsic Cu oxidation rates, Cu particles having diameters within the size range of 5 to 50 microns were oxidized with dry air in a thermogravimetric analyzer (TGA) at temperatures ranging from 200 °C to 800 °C. In the case of Cu oxidation (and CuO reduction), morphological changes in surface area, porosity, diameter and density influence observed mass loss rates and must be accounted for when extracting intrinsic kinetics from experimental data. In order to determine how these properties (i.e. surface area, porosity, diameter, and density) vary with oxidation, additional experiments were performed in a tube furnace with the same Cu particles at the same temperature / pressure and gas composition conditions as in the TGA experiments. Particles used in the tube furnace experiments were collected at different extents of conversion and analyzed to determine the parameters in the models developed to describe how surface area, diameter, porosity and density vary as a function of particle conversion. Intrinsic chemical rates were then determined by fitting a direct numerical simulation (DNS) of a single Cu particle to the TGA and tube furnace data.

The tube furnace experiments show that as the Cu particles are oxidized, pores within the particle close up, surface area decreases, and oxidation rates go from being chemically controlled to mass transport controlled. The copper oxidation mechanism employed, takes into account dissociative chemisorption of oxygen on to *Cu* and Cu_2O surfaces and *CuO* dissociation. Calculated results agree with a cascading pathway of *Cu* going to Cu_2O then to *CuO,* as experimentally observed.

(622g) Carbonaceous Fuel Conversions Through Redox Reactions of Oxygen Carrying Materials - Effect of Support

Nathan Galinsky, Yan Huang, Arya Shafie-Farhood and **Fanxing Li**, Chemical and Biomolecular Engineering, North Carolina State University, Raleigh, NC

The chemical looping strategy offers a potentially viable option for efficient carbonaceous fuel conversion with a reduced carbon footprint. In the chemical looping process, an oxygen carrying material is reduced and oxidized in a cyclic manner to convert a carbonaceous fuel into separate streams of concentrated carbon dioxide and carbon-free products such as electricity and/or hydrogen. The reactivity as well as chemical and physical stability of the oxygen carrying material are of pivotal importance to chemical looping processes.

A typical oxygen carrying material is composed of one or more multi-valence transition metal oxides supported on an "inert" supporting material. Although the supporting material does not directly contribute to the overall redox reaction to any significant extent, numerous studies have indicated that support can significantly improve the reactivity and robustness of the oxygen carrying material. In this study, the effect of support properties including surface area, melting point, and ionic and mixed-conductivity on the intrinsic redox properties of the oxygen carrying materials is investigated. An effective approach for enhancing the performance of the oxygen carrying material is offered. The tailored oxygen carrying material is shown to be active in performing redox reactions for a number of energy and environmental related applications.

(630a) Design, Analysis, Operation and Control of a Photovoltaic/PEM Fuel Cell/Nicd Battery Hybrid Renewable Energy System (HRES) for Urban Transit Applications

Zachary S. Whiteman[1], Piyush Bubna[2], Ajay K. Prasad[3] and Babatunde A. Ogunnaike[1], (1)Department of Chemical and Biomolecular Engineering, University of Delaware, Newark, DE, (2)Department of Mechanical Engineering, University of Delaware, Newark, DE, (3)Department of Mechanical Engineering, Univeristy of Delaware, Newark, DE

Design, Analysis, Operation and Control of a Photovoltaic/PEM Fuel Cell/NiCd Battery Hybrid Renewable Energy System (HRES) for Urban Transit Applications

Authors: Zachary S. Whiteman, Piyush Bubna, Ajay K. Prasad and Babatunde A. Ogunnaike

A hybrid renewable energy system (HRES), a system consisting of two or more individual renewable energy systems utilized simultaneously to meet an energy demand, potentially allows one to take advantage of complementary strengths of renewable resources in meeting energy demands in both stationary and mobile applications. Because they take advantage of multiple renewable resources, HRESs may represent a cheaper and more dependable alternative to stand-alone renewable energy systems. This presentation is concerned with a hybrid system for mobile applications, specifically, the combination of a roof-installed photovoltaic array with a PEM fuel cell/NiCd Battery bus currently operating shuttle routes on the University of Delaware campus. The primary objectives are: (i) to design the hybrid system (and an appropriate control system) to satisfy required energy demands effectively but with reduced hydrogen consumption; and (ii) to analyze the economic impact of the installation of a photovoltaic array on the PEM fuel cell/NiCd battery bus.

The study is based on actual solar irradiance data collected from a measurement device installed on the roof of the existing HRES bus which was incorporated into a Simulink model developed for the Photovoltaic/PEM fuel cell/NiCd Battery HRES. Control strategies for maintaining the desired state of charge of the NiCd battery while meeting the total energy demand of the bus were designed and implemented on the complete HRES model simulating real shuttle runs under various operating conditions. An economic analysis of the photovoltaic investment required for realizing the HRES process design was also carried out to determine the break-even period for photovoltaic array efficiencies and photovoltaic array surface areas ranging between 5% - 100% and $5m^2$ - $100m^2$ respectively. The presentation will highlight the energy management and operational performance of the control strategies as well as the economic implications of modifying the existing PEM fuel cell/NiCd HRES bus.

(630b) Issues On Energy Sustainability: Consumption, Efficiency, and Environmental Impact

Leisha Vance, Tarsha Eason and **Heriberto Cabezas**, Office of Research and Development, U.S. Environmental Protection Agency, Cincinnati, OH

One of the more important challenges in achieving sustainability is how to meet the energy consumption needs of a growing human population in the face of increasing economic prosperity. In this talk, we present a very simple but rigorous procedure for exploring the relationship between energy consumption, environmental impact, human population, gross domestic product, and energy efficiency. The underlying theory basically assumes that energy use increases with time in proportion to two factors: gross domestic product and human population. Other assumptions are that energy use has a non-zero impact on the environment and decreases only with energy efficiency. This approach allows us to pose and evaluate: (1) the limitations given the known behavior of societies and economies, practical technology, and the Laws of Thermodynamics and (2) possible pathways to sustainability. The goal is essentially to look for a way to meet expected energy consumption requirements without increasing the environmental impact beyond that seen in an appropriate reference year. We have chosen the year 1977 as our reference year because it is the year when the combined ecological footprint of the human population world-wide was equal to the biocapacity of the Earth. That is, the capacity of the Earth was able to sustain the global demand of human beings as measured by the ecological footprint. Although we use this metric as a benchmark, we are not suggesting that the ecological footprint is necessarily an ideal measure of environmental impact. Rather, we use it simply as a convenient means of establishing a reasonable reference point. Based on the aforementioned assumptions, basic calculus, and published values for energy use, gross domestic product, human population, and energy efficiency over time are used to explore energy scenarios. Because many of the sustainability issues associated with energy use, such as climate change, are global in nature, we begin with an analysis at the global scale. Given the expected growth in World gross domestic product as well as human population growth and energy use over the next 20+ years, our results indicate that the kinds of energy-use efficiencies needed to keep the environmental impact at or below the 1977 level seem unrealistic. From our calculations, energy-use efficiency would need to rise to approximately 78% by 2030 to maintain the 1977 impact level. Not only is this level nearly impossible to achieve technologically and thermodynamically, but the amount of energy consumed per unit of gross domestic product (and per capita) would need to cumulatively decrease by 1.5% to 2% per year beginning in 2012 for environmental impact of energy use not to exceed that of 1977. Again, this will not not an easy task. The situation, although quite grim, is not hopeless, and the feasibility of an integrated approach to managing the environmental impact of energy use using all of the aforementioned options will be further discussed.

(630c) Surrogate Model Based Optimal Synthesis of Solid Sorbent Carbon Capture Process

Zhihong Yuan[1], Alison Cozad[1], Nick Sahinidis[1] and David C. Miller[2], (1)Chemical Engineering, Carnegie Mellon University, Pittsburgh, PA, (2)U.S. Department Of Energy, National Energy Technology Laboratory, Morgantown, WV

Concerns about climate change have led countries to set ambitious targets for the reduction of anthropogenic greenhouse gas emissions and a number of initiatives have been set up to reduce CO2 emissions[1]. Carbon capture, utilization, and storage technology is an essential route to achieving a meaningful reduction of global CO2 emissions in the context of continued fossil fuel use in the power sector. For example, the U.S. Department of Energy (DOE) has initiated the Carbon Capture Simulation Initiative (CCSI) to accelerate the development of carbon capture technology. Solid sorbent based technologies are being used to demonstrate the capabilities of the computational tools being developed under this initiative.

Obviously, process synthesis, which has had a conspicuous impact on the development, design and operation of (petro) chemical processes[2], will play a vital role for the development of a cost-effective carbon capture process. In this study, a superstructure optimization formulation for the optimal synthesis of a carbon capture system has been established, various reactors including fluidized bed reactors and moving bed reactors together with different potential topologies have been included in the optimization formulation. Since detailed first principle models of the reactors are computationally intractable for large scale superstructure optimization, we have developed the Automated Learning of Algebraic Models for Optimization (ALAMO) approach to generate a set of low complexity algebraic surrogate models of different reactors to reduce the computational burden associated with reactor models developed in commercial process simulation packages.

Through solving the above formulated mixed integer nonlinear optimization problem which aims at minimizing the increase in cost of electricity due to capture while achieving 90% capture target the potential equipment configurations, interconnections and design/operation levels of the selected equipment can be identified. This presentation will demonstrate the application of this superstructure-based process synthesis approach for a solid sorbent-based carbon capture system.

References

1. MacDowell N, Adjiman CS, Shah N. An overview of CO2 capture technologies. *Energy Environ Sci*. 2010; 3; 1645-1669.
2. Westerberg AW. A retrospective on design and process synthesis. *Comput Chem Eng. 2004; 28: 447-458.*

(630d) Modeling of Process Steam Consumption in Chemical Batch Plants Based On Standard Production Documentation

Cecilia Pereira[1], Stavros Papadokonstantakis[2] and Konrad Hungerbühler[2], (1)Chemistry and Applied Biosciences, Institute for Chemical- and Bioengineering, Swiss Federal Institute of Technology, Zurich (ETHZ), Zürich, Switzerland, (2)Chemistry and Applied Biosciences, Institute for Chemical- and Bioengineering, Swiss Federal Institute of Technology, Zurich (ETHZ), Zurich, Switzerland

Minimization of energy consumption in the chemical industry is one of the key principles of green chemistry. Although traditionally most efforts on energy saving studies have been done for continuous production processes, batch processes have recently become the focus of many research projects [1-3]. However, a shortcut modeling approach for screening of energy consumption in multipurpose batch plants, based on a simple and general data extraction procedure from process documentation is missing. In this work a new methodology for modeling of steam consumption in multipurpose chemical batch plants based on standard process documentation, rules of thumb, expert opinion, classical thermodynamics, and some model parameterization according to Bieler et al [3] is introduced. Two case studies using production data from multipurpose batch plants of two different chemical companies were carried out for the development and validation of this new shortcut modeling approach. Several reaction and work-up processes performed in different equipments were considered for the modeling of steam consumption, and the model results were validated against reference data in each case study. Additionally, the validation results were used to derive uncertainty intervals for the model outputs based on fuzzy set theory. It has been shown that the documentation based models proposed in this work provide acceptable estimations of steam consumption in multipurpose batch plants, and that the batch-to-batch variability of the steam consumption is well captured by the uncertainty intervals resulting from the validation procedure. The possibility of allocating steam consumption to individual processes or products in multipurpose batch plants by means of this new short cut modeling approach enables the identification of energy optimization potentials.

1. Andrej Szïjjarto, S.P., * Ulrich Fischer, and Konrad Hungerbühler, *Bottom-up Modeling of the Steam Consumption in Multipurpose Chemical Batch Plants Focusing on Identification of the Optimization Potential.* Industrial & Engineering Chemistry Research, 2008. **47**: p. 7323-7334.

2. Bieler, P.S., U. Fischer, and K. Hungerbuhler, *Modeling the energy consumption of chemical batch plants - Top-down approach.* Industrial & Engineering Chemistry Research, 2003. **42**(24): p. 6135-6144.

3. Bieler, P.S., U. Fischer, and K. Hungerbuhler, *Modeling the energy consumption of chemical batch plants: Bottom-up approach.* Industrial & Engineering Chemistry Research, 2004. **43**(24): p. 7785-7795.

(630e) A Shortcut Graphical and Spreadsheet Technique for Identifying Heat Exchange Networks

Russell F. Dunn, Chemical and Biomolecular Engineering , Vanderbilt University, Nashville, TN and Kenneth A. Debelak, Chemical and Biomolecular Engineering, Vanderbilt University, Nashville, TN

Incorporating heat integration within chemical processes continues to be a critical component for future sustainability of large chemical and polymer manufacturing companies. Numerous design methodologies have been developed for the synthesis of heat exchange networks (HENs) over the past three decades. Several of the methods previously developed are based on a two stage targeting approach. The first stage uses the hot and cold composite curves to identify the maximum heat integration possible for a selected group of streams to be cooled and streams to be heated. This approach is incorporated in several commercially available software packages such as Aspen Energy Analyzer. The second stage identifies the network of heat exchangers that meets the target identified in stage one. Several strategies have been employed for identifying the HEN in stage two including, but not limited to, heuristics, mathematical optimization, and superstructure approaches.

This paper will present a new graphical/spreadsheet technique and design strategy for identifying the HEN in stage two for large industrial problems. Advantages of this strategy will be provided along with several case studies that illustrate the effectiveness of this approach. Finally, experiences using this strategy in teaching undergraduate chemical engineering design courses will also be discussed.

Sustainabilty Assessment and Design of Centralized and Decentralized Energy Solutions

Healthy economic growth in developing countries combined with increasing population and affluence are resulting in significant increase in total energy consumption. At the same time, growing concern about environmental problems such as air pollution, global warming and depletion of fresh water resources is encouraging efforts for supplying this increasing energy demand in an environmentally friendly and sustainable manner. Depletion and price fluctuation of fossil fuels also mandates finding alternative technologies to satisfy this energy demand.

Especially in rural areas of developing countries, the burden of energy poverty reduces opportunities for enhancing their quality of life. This poverty is due to factors such as absence of electrification or inconsistent availability. Utilization of inefficient energy sources like cow dung, kerosene or biomass residues causes local environmental problems such as indoor smoke further triggering health problems and related healthcare costs. Consequently, many efforts are focused on introducing decentralized and cleaner energy solutions. Such solutions are also being adopted by the developed world due to the desire for enhanced sustainability.

Given these reasons, it is essential to design energy systems consisting of centralized and localized options which create the optimum energy mix to meet energy demand in a manner that is environmentally, socially and economically feasible.

In this study, we evaluate the environmental and energetic performances of different energy technologies utilizing thermodynamic methods such as ecological cumulative exergy consumption or emergy analysis and value chain scale life cycle assessment (LCA). These technologies are two centralized coal gasification technologies (conventional and calcium looping technologies) implemented with CO_2 capture, decentralized multi-crystalline solar cells, biomass gasification, and floating drum biogas power plants. Furthermore, using a multi-objective mixed integer programming (MIP) framework and setting environmental, economic, social and thermodynamic objectives, we plan to find optimum energy mix for the selected society. This framework is applied to Rampura, a village in Central India, which is representative of a developing society.

Based on our preliminary results based on data from Rampura, biogas power plant is most attractive from a life cycle thermodynamic point of view, followed by solar power plant, biomass gasifier and with least environmental performance of coal technologies.

For rural biogas plant, we calculate its renewability index to be around 75 % for only biogas production and slightly lower for electricity generation from biogas because of machinery and maintenance requirements. A quality corrected thermodynamic return on investment is calculated to be around 5 for both biogas and electricity. A sensitivity analysis identifies manure and brick (for construction) to be the most important resources. The photovoltaic solar power plant is found to be much less attractive than the biogas plant, with the most dominant inputs being resources for open ground mounting, inverter, maintenance inputs. The renewability index of solar panel is around 2.5 % and thermodynamic return on investment is slightly over 1. Emergy analysis results of biomass gasifier unit are similar to that of solar PV, just renewability index being slightly higher. Biomass, transportation of biomass to the plant site and preparation of biomass as feedstock are the dominant inputs to the system. Not surprisingly, coal gasification is least attractive with least percentage of renewability and similar energetic performance to biomass and solar systems from an environmental point of view, but may be essential for satisfying energy needs. For coal gasification processes; mining of coal, or coal input constitutes more than half of the emergy input to the systems.

LCA enables us to capture downstream effects of processes investigated which emergy analysis cannot evaluate. According to our results, solar PV utilizes least amount of land followed by centralized coal gasification technologies and biogas plant. Likewise biogas plant is most water utilizing system followed by solar PV and coal gasification systems. While conventional coal gasification technology has higher level of GWP compared to solar PV and biogas digester technologies, CLP has lowest GWP. This situation points out that improved coal gasification technologies implemented with CO_2 capture can significantly ease environmental burden of generating electricity from coal.

Life cycle assessment of biomass gasifier plant is the next steps in this work followed by the formulation of the optimization problem to identify the "optimum" energy mix while accounting for multiple objectives. This framework will be relevant to choosing combinations of centralized and decentralized options in developing and developed economies.

"Challenges and Opportunities for Enhancing Exergy Efficiency of Industrial Sectors in Developing Economies – Insight From India's Growth"

Sachin B. Jadhao1, Aniruddha B. Pandit1 and Bhavik R. Bakshi2,

1. Institute of Chemical Technology, Mumbai, India

2. William G. Lowrie Department of Chemical and Biomolecular Engineering The Ohio State University, Columbus, OH.

Abstract

Given that the development of emerging economies is invariably accompanied by an increase in energy and resource consumption, goals of sustainable development require a parallel increase in resource efficiency. This talk will focus on the trends and opportunities for enhancing the energy and material efficiency of industrial sectors in India. It uses the approach of exergy analysis since it has been found to be an appropriate measure for resource accounting. Studies on industrial sectors have been carried out for several economies and the resulting insight can be used for identifying opportunities for enhancing resource use and reducing environmental impact. The aim of this work is to understand the metabolism of Industry sector of India as it has evolved from a controlled to a more open free market economy, and its implications for sustainable development. We analyzed various industries in India on the basis of, resource consumption and environmental pollution. We estimate second law efficiencies for various industries in India, including Chemical, Iron and steel, Fertilizer, cement, sugar and others. We also estimate the efficiency of the India's industry sector as a whole. Trends of resource consumption and its conversion into useful work are analyzed using exergy analysis and the results are compared with those of other developing and developed economies. In addition, using techniques of network analysis, opportunities for enhancing exergy efficiency are identified.

Introduction

Industry sector in India is one of the major resource consuming sector. Figure 1 and figure 2 shows the fuel exergy and electric exergy consumed in Industry sector of India since 1970. The physical resources (fuel and non fuel minerals) consumption in Industry sector in India has been observed to increasing rapidly from the period of start if industrialization and economic liberalization in India. Among the various industries Inron and steel, fertilizer, Chemical and Cement are the major resource consuming industries. In general the industries are evaluated on the basis of First Law efficiency (calculated by comparing the energy content of the useful output with the energy content of the inputs, including materials), however the first law does not take in to account the quality which leds to the misleading conclusions [1].

Exergy efficiency is the ratio of potentially useful (exergy) output to potentially useful (exergy) input. [1] Second-law efficiency is usually less than first law efficiency.[1] Exergy efficiency captures the degree to which energy inputs are converted into useful work [3]. Exergy analysis provides insights not available via other methods, especially for identifying unit operations with the greatest potential for improvement [2].

A number of efforts have been undertaken to apply exergy analysis to various sectrors in the economy that includes the industrial sector. Since Industry sector accounts for major share in total resource consumption of many economies (particularly with respect to developing economies like India and China), it is important to examine the efficiency of resource utilization in this sector. Many studies in past have analysed the Industrial sector of various economies on the basis of exergy efficiency. These studies presented exergy accounts for individual process stages, in order to identify major losses and evaluate the potential for further technical improvements in the processes. Industrial sector in India have been compared and analyzed on the basis of efficiency or resource conversion in different industries. Most of the existing research on thermodynamic analysis of industry mainly accounts for fossil fuel consumption and its conversion to useful work. Few studies have been conducted using exergy assessment of industries that takes in to account fuel, non-fuel minerals resources. In this study we have analyzed the industries on the basis of fuel and non-fuel resource consumption and emissions.

Research method

In this work, conversion and utilization of all natural resources including energy and materials are described in terms of exergy flow in various Industries in India. Exergy utilization trends and efficiencies for various industries are analyzed and compared with those of other studies. Exergy losses in transformation of material resources and in conversion of various forms of energy into useful work are described in detail.

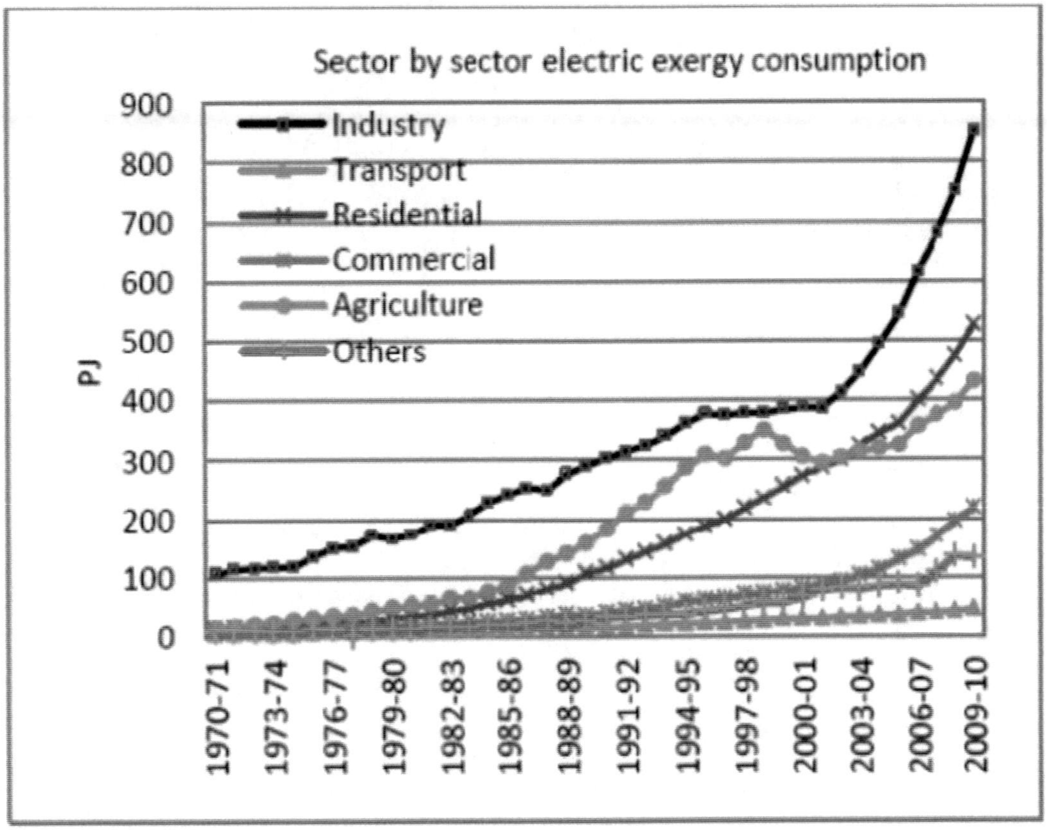

Figure 1 Electric exergy consumed in Industry sector of India

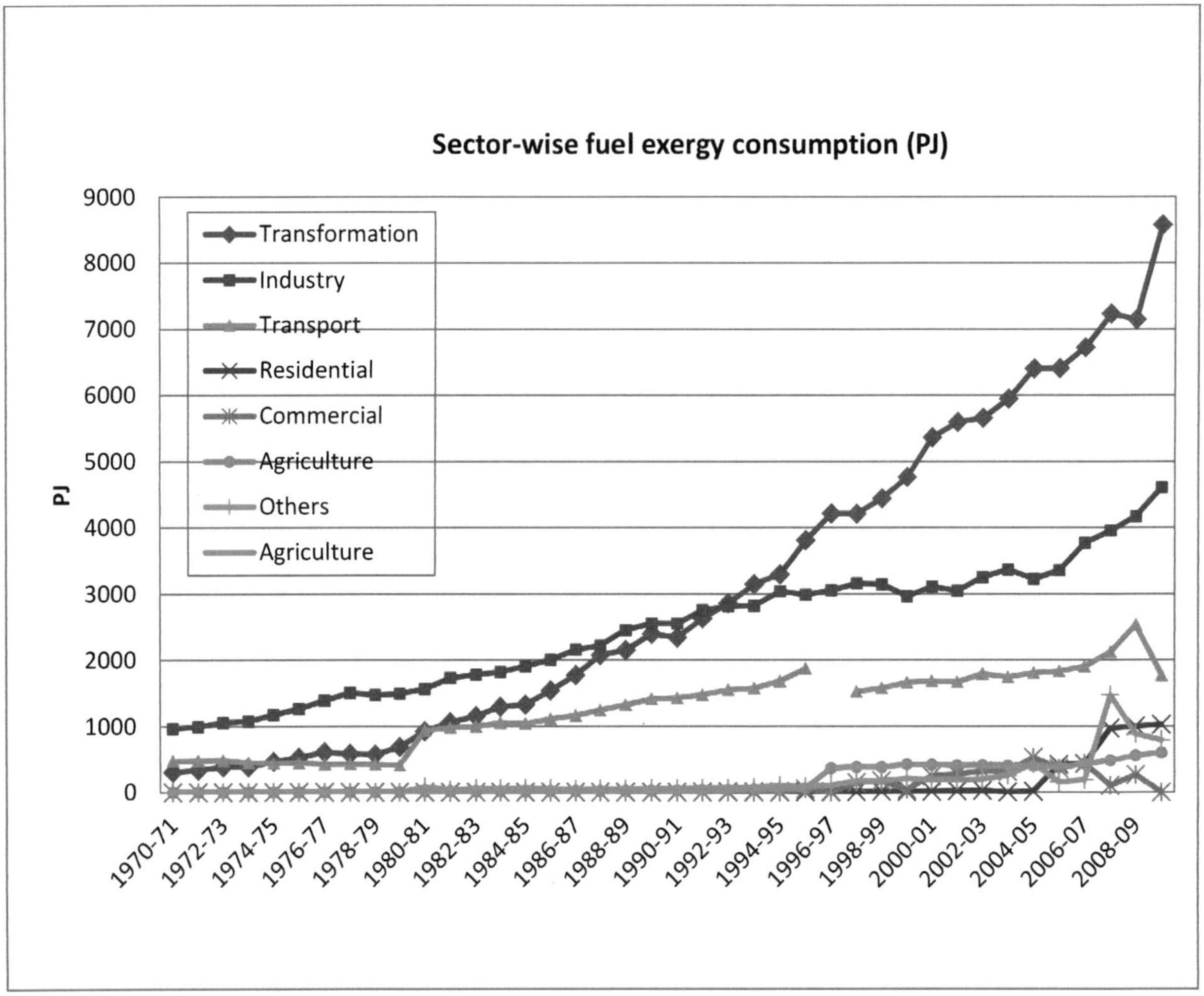

Figure 1 Fuel exergy consumed in Industry sector of India

Sample calculation of exergy efficiency for Iron & Steel Industry in India (2007-08)

Energy utilization in the industry sector is categorized according to the energy use as follows

Energy use in the industrial sector in India by type of Use (%)		
Use Category	Electrical	All Fuels
Mechanical Drive	65	0
Heating	28	100
Others (Assumed Lighting)	7	0
TOTAL	100	100

INPUTS

Total Electric energy consumed in the Iron & Steel Industry= 75.78 **PJ**

Total Electric exergy consumed in the Iron & Steel Industry= 75.78 **PJ**

Total Fuel energy consumed in the Iron & Steel Industry= 369.21 **PJ**

Total Fuel exergy consumed in the Iron & Steel Industry= 391.37 **PJ**

[I] Mechanical Drive

Since 65 % of the electric energy is used for mechanical drive application

$$Energy\ \&\ exergy\ input\ for\ machenical\ drive = 75.78 \times 0.65 = 49.26\ PJ$$

Mechanical drives are assumed to be 90 % efficient

$$Enegy\ \&\ exergy\ output\ for\ machenical\ drive = 49.26 \times 0.90 = 44.33\ PJ$$

[II] Heating

Process heating in the industry is subdivided in to three categories based on the temperature ranges as follows. The temperature range and the efficiency values are taken from the study of Dincer et al.

Process heating temperatures and efficiencies for the industrial sector			
Category	Temperature	Electrical	Fuel

	(° C)	(%)	(%)
Low	<121	100	65.5
Medium	121–399	90	60
High	>399	70	50

Breakdown of the energy used for each temperature range in the industry is referred from the Brown et al. study for 108 industrial processes

From this study the breakdown for Iron & steel Industry is given as follows

Breakdown of energy used for each temperature (%)					
Industry	T range	TO	Mean T (°C)	Electricity	Fuel
Iron and steel	Low	25	45	4.2	0
	High	25	983	95.8	100

Exergy efficiency of low temperature electric heating = 45+273 = 318°K and **η= 100 %**

$$\Psi_{e\,h\,Low} = 1 \times \left[1 - \left(\frac{298}{318}\right)\right] = 6.28\ \%$$

Exergy efficiency of high temperature electric heating: T = 983+273 = 1256 °K and **η= 70 %**

$$\Psi_{e\,h\,high} = 0.7 \times \left[1 - \left(\frac{298}{1256}\right)\right] = 53.39\%$$

Exergy efficiency of low temperature fuel heating: T = 45+273 = 318 °K and **η = 65 %**

$$\Psi_{f\,h\,Low} = 0.65 \times \left[1 - \left(\frac{298}{318}\right)\right] = 4.08\ \%$$

Exergy efficiency of high temperature fuel heating: T = 983+273 = 1256 °K and **η = 50 %**

$$\Psi_{f\,h\,high} = 0.5 \times \left[1 - \left(\frac{298}{1256}\right)\right] = 38.13\ \%$$

Energy efficiency of electric heating= efficiency×percentage of electric energy used

$$\eta_e = 100\%\times 4.2\% + 90\%\times 0\% + 70\%\times 95.8\% = 71.26\%$$

Energy efficiency of fuel heating= efficiency× percentage of fuel energy used in iron & steel industry

η_f= 65%× 0%+ 60%×0%+50%×100% = 50%

Exergy efficiency of electric heating= efficiency×% electric energy used in iron & steel industry

Ψ_e= 6.28%× 4.2%+ 53.39%×95.8% = 51.0%

Exergy efficiency of fuel heating= efficiency×% fuel energy used in iron & steel industry

Ψ_f= 7.2%× 0%+ 38.7%×100% = 38.7%

In Industry 28 % of the electricity is used for heating applications.

Energy Efficiency of heating

$$\eta_q = \frac{(Electric\ energy\ used \times efficiency + Fuel\ energy\ used\ \times efficienfy)}{(Electric\ energy\ + Fuel\ Energy\ input)}$$

$$\eta_q = \frac{(75.78 \times 0.28 \times 0.713 + 369.21\ \times 0.50)}{(75.78 \times 0.28\ + 369.21)}$$

$$\eta_q = 51.15\ \%$$

Exergy Efficiency of heating

$$\Psi_q = \frac{(Electric\ exergy\ input \times efficiency + Fuel\ exergy\ input\ \times efficienfy)}{(Electric\ exergy\ + Fuel\ Exergy\ input)}$$

$$\Psi_q = \frac{(75.78 \times 0.28 \times 0.514 + 391.37\ \times 0.387)}{(75.78 \times 0.28\ + 391.37)}$$

$$\Psi_q = 39.35\ \%$$

$$Energy\ Output\ of\ heating = Energy\ Input \times Energy\ Efficiency\ for\ heating$$

$$Energy\ Output\ of\ heating = 75.78 \times 0.28 \times 0.72 + 369.21 \times 0.50 = 199.88\ PJ$$

$$Exergy\ Output\ of\ heating = Exergy\ Input \times Exergy\ Efficiency\ for\ heating$$

$$Exergy\ Output\ of\ heating = 75.78 \times 0.28 \times 0.52 + 391.37 \times 0.387 = 162.35\ PJ$$

[III] Lighting

7 % of electricity is used for lighting application

$$Energy\ \&\ exergy\ input\ for\ lighting = 75.78 \times 0.07 = 5.30\ PJ$$

Lighting energy and exergy efficiencies are assumed as **18.5% and 17.1%**

$$Enegy\ output\ for\ lighting = 5.30 \times 0.185 = 0.98\ PJ$$

$$Exegy\ output\ for\ lighting = 5.30 \times 0.17 = 0.90\ PJ$$

$$Total\ Energy\ Input = Aggrigate\ energy\ input\ for\ heating, lightening\ and\ mechanical\ drive$$

$$Total\ Energy\ Input = 369.21 + 75.78 = 444.99\ PJ$$

$$Total\ Exergy\ Input = 391.37 + 75.78 = 466.78\ PJ$$

$$Total\ Energy\ output = 199.88 + 44.33 + 0.98 = 245.19\ PJ$$

$$Total\ exergy\ output = 162.35 + 44.33 + 0.90 = 207.58\ PJ$$

$$\boldsymbol{Overall\ energy\ efficiency} = \frac{Total\ energy\ output}{Total\ energy\ input} \times 100$$

$$= \frac{245.19}{444.99} \times 100 = \mathbf{55.10}\ \%$$

$$\boldsymbol{Overall\ exergy\ efficiency} = \frac{Total\ exergy\ output}{Total\ exergy\ input} \times 100$$

$$= \frac{207.58}{466.78} \times 100 = \mathbf{44.34}\ \%$$

Results and discussion

This work provides insight into areas where major improvements are required to enhance resource efficiency in the various industries. Effects of factors such as increasing consumption and depletion of resource quality on overall efficiency are demonstrated.

REFERENCES

[1] R. U. Ayres, L. T. Peir, and G. V. Mendez. Exergy e-fficiency in industry where do we stand? Environ. Sci. Technol, 2011.

[2] B. R. Bakshi and G. F. Grubb. Appreciating the role of thermodynamics in lca improvement analysis via an application to titanium dioxide nanoparticles. *Environmental Science and Technology,* 45, 7, 3054–3061, 2011.

[3] E. Williams, B. Warr, and R. U. Ayres. Efficiency dilution: Long-term exergy conversion trends in japan. Environ. Sci. Technol., 42:49644970, 2008.

(636a) Fundamentals of Electrochemical Energy Storage

Sri Narayan, Chemistry, University of Southern California, Los Angeles, CA

Electrochemical energy storage systems are used to efficiently store and transport electrical energy. The applications for such systems range from portable electronics to vehicles and large-scale power plants. This tutorial will describe the operating principles, materials requirements, rate processes and the design approaches relating it to the requirements of the applications. The tutorial will discuss the features and performance characteristics of metal-hydride batteries, lithium-ion batteries, metal-air and metal-sulfur batteries, and redox flow type batteries. The technical challenges to designing inexpensive, eco-friendly, and robust electrochemical energy storage systems and the recent technological advances will be discussed.

(636b) Materials Challenges of Electrochemical Energy Storage Technologies

Arumugam Manthiram, Materials Science and Engineering, University of Texas at Austin, Austin, TX

Electrical energy storage (EES) is critical for the efficient utilization of renewable energies like solar and wind as well as for electrifying the transportation sector. Rechargeable batteries and supercapacitors are the primary candidates for EES. Cost, safety, cycle life, energy, power, and environmental impact are the major issues hampering the adoption of these technologies. This presentation will first provide an overview of the basic principles involved in electrochemical energy storage, followed by an account and status of the existing electrode materials for lithium-ion batteries. After pointing out the materials challenges and opportunities, the presentation will then focus on (i) next generation of cathode and anode materials for lithium-ion batteries, (ii) high energy, high power lithium-sulfur batteries, and (iii) high-energy density dual-electrolyte lithium-air cells employing a lithium-ion conducting solid electrolyte. Particularly, the importance of novel synthesis approaches, surface modification, electrode architecture, and optimum cell configurations to realize the necessary performance parameters will be emphasized.

(636c) Promise and Challenges of Materials for Electrochemical Capacitors

Prashant Kumta, Bioengineering, University of Pittsburgh, Pittsburgh, PA

Nanotechnology is a revolutionary area that has impacted several areas of materials science and technology including the area of electrochemically active materials for lithium-ion batteries, electro-active catalysts for fuel cell systems, and electrochemical capacitors. The growing worldwide demand for energy makes the need for sustainable energy and power storage systems highly critical. In this regard electrochemical supercapacitors are emerging as devices of prime importance owing to their superior characteristics unparalleled by any other charge storage device. These include: high power densities at relatively high energy densities and long cycle life. Electrochemical capacitors comprise electrochemical double layer (EDLC), pseudo-capacitors, and supercapacitors. All of these systems are differentiated by their intrinsic distinct fundamental mechanisms and are all capable of delivering large bursts of charge over a very short period of time. It is due to this unique capability that electrochemical capacitors find a very important role in the growing alternative energy field. High surface area carbons, hydrated ruthenium oxide, nanostructured VO_x and MnO_2, high surface area and highly electronically conducting graphene, carbon nanotubes (CNTs) are among the primary materials being considered for supercapacitor applications. In recent years there has also been significant activity in transition metal non-oxide based electrochemical capacitors. This presentation will cover the fundamental principles of electrochemical capacitors and discuss the current materials of choice. The presentation will culminate by providing insights into the challenges and promise of novel next generation materials for use in electrochemical capacitors for high power applications.

(636d) Fundamentals for Designing Flow Batteries

J.W. Van Zee, Karson Leperi and Victor Caldwell, Chemical Engineering, University of South Carolina, Columbia, SC

Selected flow battery systems will be used to present the fundamentals of and advances in electrochemical engineering design. Applications of these design principles will allow assessment of performance, degradation, and opportunities for various load cycles. The interaction of economics with the principals of thermodynamics, kinetics, mass transfer, and potential distributions will be discussed. In the spirit of a tutorial, analogies will be drawn between common chemical engineering concepts and electrochemical engineering principles. The goal is to provide the basis for an understanding of the limitations and the advantages of flow batteries. Examples of the use of experimental designs to isolate limiting behavior and opportunities for improvement will be discussed. Opportunities for improvements using commercial computational fluid dynamics packages will also be discussed.

The fundamentals are introduced with a simple algebraic model for a hybrid flow battery and stack that uses zinc-bromine chemistry. Performance is shown to be lost though subtractions from the thermodynamic voltage by ohmic losses which can be a function of geometry, by columbic efficiency losses which can be a function of separator design, and by voltage efficiency losses which are related to kinetic, mass transfer in the cell, and shunt currents in the stack. A review of more sophisticated models and the opportunities for cell and stack improvement will be discussed.

Then the paper will discuss the necessary concepts for extension to a complete flow battery with an example of replacing the zinc negative electrode with hydrogen. Novel designs and limitations are discussed for this system with a brief review of porous electrode theory. The economics of the system are discussed components as well as the interactions of the effects of geometry, component selection, and auxiliary power and energy requirements.

(636e) Quantifying Solute-Solute Interactions in Flow-Battery-Membrane Transport

Lucas Griffith, Chemical Engineering, University of Michigan, Ann Arbor, MI and Charles W. Monroe, Department of Chemical Engineering, University of Michigan, Ann Arbor, MI

Redox Flow Batteries (RFBs) have the potential to provide the high-capacity energy storage needed to improve power grid efficiency through peak shaving, as well as aid in the integration of renewable energy sources, such as wind energy and solar energy. The high cost of traditional ion-selective membranes has motivated the exploration of the many different membranes available for use in RFBs[1]. In order to facilitate this exploration it is necessary to characterize mass transport in various membranes, so that the effect of using different membranes on RFB performance can be modeled.

Although treatment of binary diffusion in all-vanadium RFBs is ubiquitous in the literature[2], there is a surprising lack of effort to describe the effect of ion-ion interactions during multicomponent transport across ion-selective membranes. Heintz *et al*[3] have investigated these effects for sodium and hydrogen diffusing across Neosepta membranes and have seen that the magnitude of the ion-ion interactions is comparable with the magnitude of the ion-solvent interactions. This shows that the importance of including ion-ion interactions in models of RFBs should be tested, so we have measured ion-ion interactions for $VOSO_4$ and H_2SO_4in two different membranes.

This talk will briefly discuss how Barnes's transient diffusion model of dialysis cells[4] can be solved and used to measure binary diffusion coefficients and sorption equilibria simultaneously. Results of binary experiments can be used as a basis for describing ternary interdiffusion measurements, which can be modeled by applying the Stefan-Maxwell transport formalism. For n diffusing species, this system of n force-explicit equations can be reduced to a system of n-1 flux explicit equations by taking the membrane velocity as the reference and applying the Gibbs-Duhem equation[5]. Taylor and Krishna[6] have shown that a batch dialysis cell will decay exponentially toward equilibrium at long times, with decay constants that are the eigenvalues of the system of transport equations. Therefore, if the binary diffusion coefficients are known, solute-solute interactions can be measured with an interdiffusion experiment in a batch dialysis cell. The values of the binary diffusion coefficients as well as the solute-solute interaction parameters will be given for $VOSO_4$ and H_2SO_4 (species relevant to aqueous all-vanadium RFBs) in two exemplary hydrated membranes.

1. Skyllas-Kazacos, M., M. H. Chakrabarti, et al. (2011). J. Eletrochem Soc **158**(8): R55-R79

2. Wiedemann, E., A. Heintz, et al. (1998). J. Membrane Sci. **141**(2): 215-221.

3. Heintz, A., E. Wiedemann, et al. (1997). J. Membrane Sci. **137**(1-2): 121-132.

4. Barnes. (1934). Physics **5**(1): 5.

5. Meyers, J. P. and J. Newman (2002). J. Electrochem. Soc. **149**(6): A718-A728.

6. Taylor, R. and R. Krishna (1993). Multicomponent mass transfer, Wiley.

(637a) Gas-Phase Visible Light Photocatalytic Activity of Flame Aerosol Synthesized Nanostructured Transition Metal Modified TiO2

siva Nagi Reddy Inturi, Chemical Engineering Program, School of Energy, Environmental, Biological & Medical Engineering, University of Cincinnati, cincinnati, OH, Makram Suidan, american university of beirut, Beirut, Lebanon and Panagiotis Smirniotis, Chemical Engineering Program, School of Energy, Environmental, Biological and Medicinal Engineering, University of Cincinnati, Cincinnati, OH

In the recent year there is an increasing interest in developing visible light photocatalytic degradation of volatile organic compounds. In our present studies, a series of transition metal (Cr, V, Fe, Cu, Mn, Co, Ni, Mo, and La) incorporated TiO_2 materials have been synthesized by using a flame aerosol synthesis technique. The resulting materials were explored for the destruction of VOCs in the presence of visible as well as UV light. A combination of various physico-chemical characterizations such as N_2 physisorption, O_2 chemisorption, X-ray diffraction (XRD), diffuse reflectance UV-vis (DR UV-vis), and temperature program reduction (TPR) were used to characterize the chemical environment of these transition metals in as-prepared catalysts. The dispersion of transition metals as determined by O_2 chemisorption suggests that they are well dispersed on TiO_2. The UV-vis spectroscopy results of M-TiO_2 materials (where M= Cr, V, Fe, Cu, Mn, Co, Ni, Mo, and La) showed enhancement of light absorption in the visible range by selected composite materials in the UV range (250-400 nm) due to the presence of titania. Among all of the catalysts, the transition metals (M=Cr and Fe) incorporated materials have shown a very promising catalytic performances in visible light. Their catalytic performances are correlated with the UV-vis spectrum of each synthesized catalyst to reveal the specific role played by each metal ion. The photocatalytic performance of these flame aerosol synthesized transition metal incorporated TiO_2 material will be presented in this paper.

Quantum Yield in a Photo-CREC reactor for Hydrogen Production

Salvador Escobedo[a], Benito Serrano[b], Hugo de Lasa[a,*]

[a] *Western University, London, N6A 3K7, Canada*
[b] *Universidad Autonoma de Zacatecas, Zacatecas, 98000, Mexico*
**Corresponding author: hdelasa@eng.uwo.ca*

Introduction

Nowadays, photoelectrodes and photocatalysts are being developed for water splitting to produce hydrogen. The goal of this study is to demonstrate the feasibility of water splitting for hydrogen production through a Photo-CREC reactor, using a sacrificial agent and a modified Platinum loaded semiconductor of DP25 (TiO_2) [1] It is also the objective of this study determining optimum Pt loadings, as well as quantum yields [2 and 3].

Experimental

Degussa P25 (TiO_2) was impregnated with chloroplatinic acid. This step was followed by calcinations and reduction. A 2.73 eV reduced band gap was observed for the modified semiconductor using UV-Vis NIR spectrophotometer (see Figure 1). The prepared Pt doped photocatalysts were studied in a Photo-CREC-Water II. Macroscopic irradiation balances were performed for establishing absorbed irradiated energy (see Figure 2) and quantum yields.

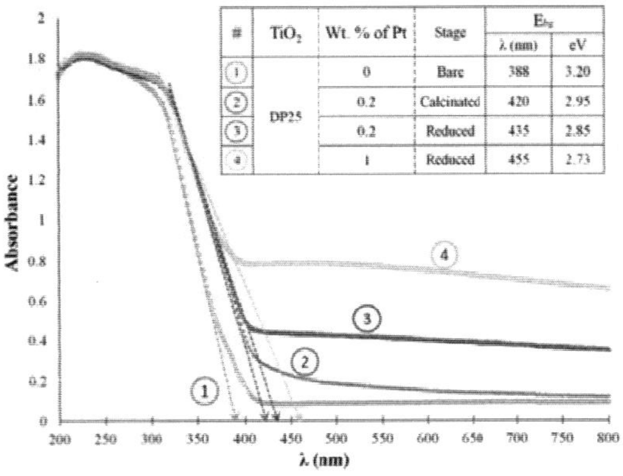

Figure 1. UV-Vis Band Gap

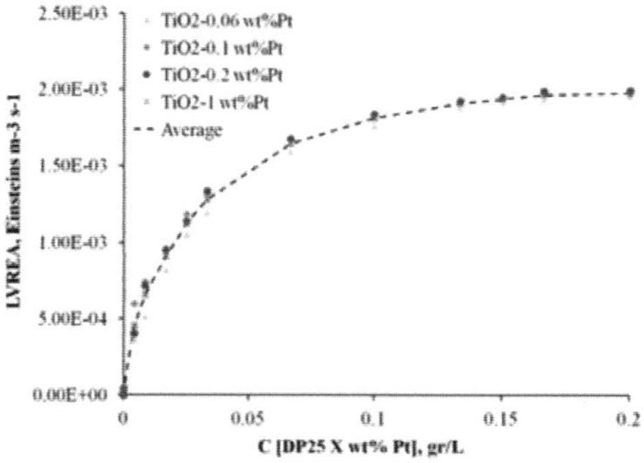

Figure 2. LVREA within the Photo-CREC reactor using a macroscopic balance

Results and Discussion

Impregnation of Pt on TiO_2 leads to enhanced particle size distribution with reduction of particle agglomeration (see Figure 3), with better distribution of charges on the semiconductor. On the other hand, it is shown that Pt did not affect the anatase and rutile phases as well as the specific surface area of the DP25. Figure 4 reports hydrogen production as a function of contact time under free oxygen conditions and pH equals to 4. Runs were developed with excess of Ethanol as $OH^{\bullet}$ scavenger reagent. Hydrogen free of oxygen was produced in all cases with 1wt% Pt on Degussa P25 yielding the best hydrogen production rates. Figure 5 reports quantum yields in a 6-liter Photo-CREC reactor with Ethanol at 2% v/v concentration and different Pt loads on Degussa P25 (see Table 1). It is proven that Ethanol helps averting h^+ and e^- recombination, with the highest 7.86% quantum yields obtained between 1-6 hours of irradiation.

Table 1 Reaction Rate at different Loading of Pt on TiO_2 and pH=4

Catalyst	Reaction rate [a] (mol h^{-1} g_{cat}^{-1})	Quantum Yield (%)
		H_2 Production
DP25 0.06wt% Pt	88.92	1.82
DP25 0.1wt% Pt	142.26	2.92
DP25 0.2wt% Pt	200.06	4.12
DP25 1wt% Pt	382.97	7.86

[a] Reaction conditions: 298 K, 1 atm.

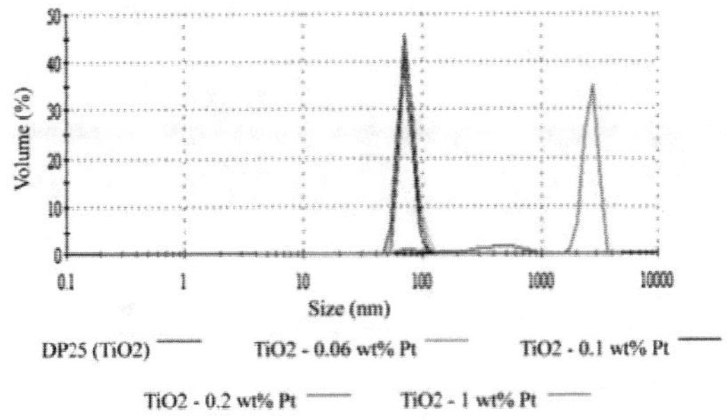

Figure 3. Particle Size Distribution among different Catalyst

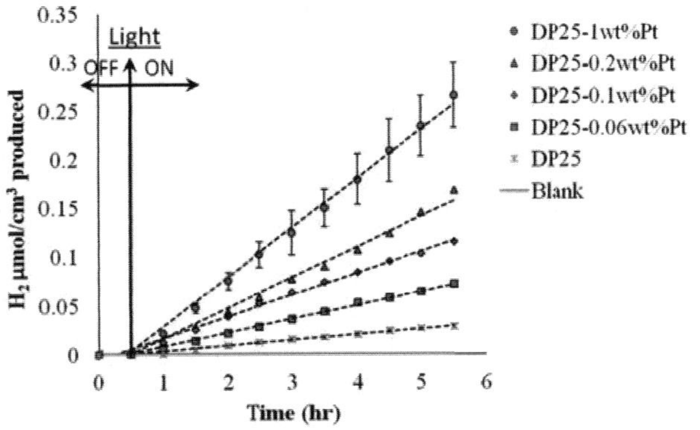

Figure 4. Hydrogen Production under pH=4, Ethanol 2%v/v and Atmosphere of Argon.

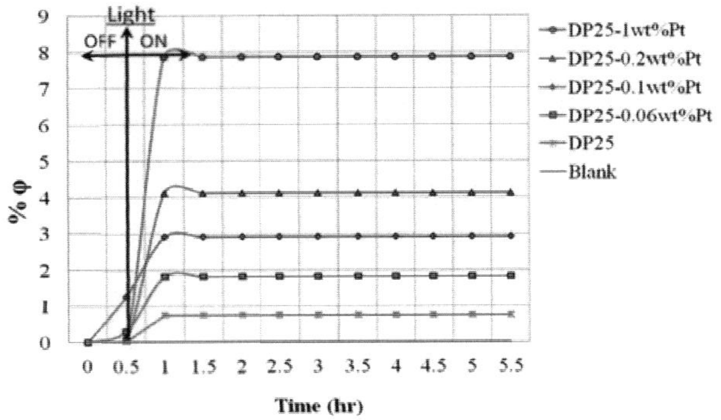

Figure 5. Hydrogen Quantum Yield in different catalyst

References

1. P. Alexia, K. Dimitris I. et al. Cat. today. 124 (2007) 94

2. B. Serrano, A. Ortiz, J. Moreira and H. de Lasa, *Ind. Eng. Chem. Res.*48 (2009) 9864

3. De Lasa Hugo, Serrano Benito and Salaices Miguel, "Photocatalytic Reactor Engineering", *Springer Science* (2005), pp 187; ISBN 0-387-23450-0

(637c) Low-Temperature, Manganese Oxide-Based, Thermochemical Water Splitting Cycle

Bingjun Xu, Chemical Engineering, Caltech, Pasadena, CA and Mark E. Davis, Chemical Engineering, California Institute of Technology, Pasadena, CA

Thermochemical cycles that split water into stoichiometric amounts of hydrogen and oxygen below 1000°C are highly desirable, because they can convert the waste heat from nuclear power plants into chemical energy in the form of hydrogen. We report a manganese-based thermochemical cycle with the highest operating temperature of 850°C that is completely recyclable and does not involve toxic or corrosive components. The thermochemical cycle includes the following 4 steps: (1) thermal treatment of a physical mixture of Na_2CO_3 and Mn_3O_4 to produce MnO, CO_2, and alpha-$NaMnO_2$ at 850°C, (2) oxidation of MnO in the presence of Na_2CO_3 by water to produce H_2, CO_2 and alpha-$NaMnO_2$ at 850°C, (3) Na^+ extraction from alpha-$NaMnO_2$ by its suspension in aqueous solutions in the presence of bubbling CO_2 at 80°C, and (4) recovery of Mn_3O_4 by thermally reducing the sodium ion extracted solid produced in step (3) at 850°C. The shuttling of Na^+ into and out of the manganese oxides in the hydrogen and oxygen evolution steps, respectively, provides the key thermodynamic driving forces, and allows for the cycle to be closed at temperatures below 1000°C. The production of hydrogen and oxygen is fully reproducible for at least 5 cycles.

(637d) Pyrite Oxidation in a Packed Bed Configuration

Edward Trujillo, Chemical Engineering, University of Utah, Salt Lake City, UT and Surya P. Sunkavalli, Chemical Engineering, university of Utah, Salt Lake City, UT

Pyrite is one of the more common minerals found in mining operations and is one of the major minerals involved in the formation of acid rock drainage. The oxidation and the formation of acid rock drainage is a very complex process occurring in rock piles that depends on oxygen levels, temperature, pH, redox potentials, ionic concentrations and bacteria concentrations. Abiotic oxidation of pyrite can occur with oxygen or with ferric ion and is strictly chemical but biotic oxidation requires the presence of iron-oxidizing and/or sulfur-oxidizing bacteria, the two more common species being *acidothiobacillus ferrooxidans* and *leptospirillum ferrooxidans*. Despite numerous studies and publications, the mechanism of pyrite oxidation, both abiotic and biotic, is still somewhat controversial and terms such as "direct" and "indirect" oxidation have been used to describe the process. This paper discusses the oxidation of pyrite under controlled conditions in a packed bed configuration. Experiments were conducted with pyrite crystals in a packed bed of glass beads over time periods of several weeks and results showed that oxidation rates varied with leaching pH and composition. Acidic conditions with 100 mg/L Fe^{+3} under water saturated conditions displayed the most rapid rates.

(637e) Experimental and Statistical Optimization of Chelant Assisted Extraction of Heavy Metals From Bimetallic Spent Catalyst

Garima Chauhan, Chemical Engineering, Indian Institute of Technology Delhi,, New Delhi, India, K.K. Pant, Chemical Engineering Department, Indian Institute of Technology, Delhi, New Delhi, India and K. D. P. Nigam, Department of Chemical Engineering, Indian Institute of Technology Delhi,, New Delhi, India

Extraction of heavy metals from the spent catalyst is of great concern these days to overcome the scarcity of available resources and to minimize the metallic waste. Disposal of metallic waste directly to the environment may cause severe health issues due to environmental pollution. Conventional technologies viz. leaching, roasting, smelting etc., which been have been developed for the recovery of heavy metals, are not environmental friendly.

In the present study, an ecofriendly approach has been adopted for the extraction of heavy metals from bimetallic spent catalyst (CoMox) of a hydroprocessing unit. The catalyst procured from secondary reforming unit contained Mo (11-13%), Co (2-3%), with the balance being the support material γ-Al_2O_3. A chelating agent ethylene diamine tetraacetic acid (EDTA) was employed for the extraction of Co and Mo. All the experiments were performed at atmospheric pressure under reflux condition and reaction parameters (temperature, solid to liquid ratio, time, molar ratio of EDTA and metal, pH, stirring speed, and particle size) were optimized to maximize the extraction. Cobalt was separated in form of precipitate at alkaline pH while molybdenum was recovered in form of MoS_2at pH 3. Direct and derivative spectrophotometric determination of molybdenum (VI) was performed using cinnamaldehyde-4-hydroxy benzoylhydrazone (CHBH). 85% molybdenum and 76% cobalt was extracted from spent catalyst at optimum reaction conditions. Characterization of spent catalyst, residue and precipitated metals was also performed to understand the structural and phyisco-chemical properties. Residue contains a large amount of alumina. This alumina content was reused as a support in the preparation of other catalyst.

The effect of process parameters on the extraction of Co and Mo were quantitavely evaluated by the analysis of variance (ANOVA). Box-behnken design with various process parameters at three different levels (-1,0,1) was coupled with response surface methodology for designing of experiment. Statistical analysis of the results showed that the linear and quadric terms of the variables had significant effects to the response. However, interactions terms did not show a substantial impression on Co extraction. 3D response surface plots and corresponding 2D contours were examined to study the effect of process parameters on extraction of Ni. The optimal conditions of process parameters were predicted by the proposed models. Verification of the optimization showed a small percentage error between the predicted and the experimental values of Ni extraction.

The process is significant demonstration of the use of chelation chemistry for metal extraction from spent catalyst. Use of sulphuric acid for dechelation and metal precipitation is limited to low temperatures making it a non-corrosive and energy-efficient process. Bimetallic extraction from spent catalyst and recoverability of EDTA make it an eco-friendly approach. Efforts are still ongoing to setup a process with zero waste generation. Statistical optimization of process parameters for extraction of Ni using the box-behnken design appeared to be a valuable tool for the extraction of Ni. It has been demonstrated that box-behnken design coupled with response surface methodology allows the optimization of the experimental variables.

(637f) Formation of Disinfection by-Products in Ballast Water Chlorination Process

Yue Ma, Department of Civil and Environmental Engineering, National University of Singapore, Singapore, Singapore and J. Paul Chen, Department of Civil & Environmental Engineering, National University of Singapore, Singapore, Singapore

Formation of Disinfection By-products in Ballast Water Chlorination Process

Yue Ma, J. Paul Chen*

Department of Civil and Environmental Engineering

National University of Singapore, 10 Kent Ridge Crescent, 119260 Singapore

* Corresponding author. Tel.: +65 6516 8092; Fax: +65 6872 5483, +1 831 303 8636. *Email addresses:* jchen.enve97@gtalumni.org, ceecjp@nus.edu.sg(J.P. Chen).

Abstract

Chlorination has been proved as an effective option for minimizing and preventing the introduction of invasive marine species from ballast water. However, the formation of disinfection by-products (DBPs) would be a potential risk associated with the usage of chlorine as a biocide for ballast water treatment. The objective of this research is to access the formation potential of by-products, including four trihalomethanes (THMs) and nine haloacetic acids (HAAs) in seawater sample under different chlorination conditions likely to be encountered in ballast water disinfection system. The effects of chlorine concentrations, total organic carbon concentration and operation temperature on formation potential of DBPs were investigated. The decrease in UV absorbance at 254 nm (ΔA_{254}) and chlorine decay in the chlorination process were also monitored. It was determined that THMs predominated by bromoform while BDAA was the main HAAs component in all the samples studied. High fractions of bromo-deviation formed could be due to high amounts of bromide in seawater which is rapidly transformed by chlorine to hypobromous acid and other forms of bromine. Although the occurrence of DBPs in the investigated samples was higher than regulation amounts for treated drinking water, the environmental risk assessment showed the treated water would not pose great threats to the marine environment.

Keywords: By-products; Chlorination; Ballast water; Kinetics

ID# 254226:

Experimental study of the operating parameters and biodegradation kinetic model for wastewater treatment in a biological fluidized bed reactor

Kun Hu, Weixing Huang, Zhenzhong Li, Kaiyuan Niu, Mengli Yu

School of Chemical Engineering, Sichuan University, Chengdu, Sichuan, 610065, P. R. China.

Abstract

The effects of the wastewater contents, air supply, reaction temperature and pH on the treatment performance were investigated through systematic experiments in a three-phase biological fluidized bed rector with mature biofilm cultivated on synthetic porous particles made of poly (MA-VAc-MMA). Synthetic wastewater consisted glucose, NH_3Cl, KH_2PO_4 and modestly additional nutrient salts was applied in the experiments. The results indicated that the initial COD and NH_3-N concentrations as well as the COD/NH_3-N ratio in the wastewater had impacts on their removals. The optimal treatment performances were gained with COD of 700 mg/L, NH_3-N of 25 mg/L and the COD/NH_3-N ratio of 100/4~100/3 in the raw wastewater, respectively. The variations of air flow rate affected both the dissolved oxygen concentration in the reactor and the shear force on the surface of the biofilm-coated particles, and thus, there existed an optimal air flow rate at which the most effective treatment was achieved. It was also found that the favorable range of temperature and pH for the biodegradation process were 20~25℃ and 6~8 respectively, and beyond which the treatment performance would be obviously inhibited. Meanwhile, the semi-continuous experiments were conducted to study the dynamic behaviors of COD and NH_3-N removals in the reactor, and the response curves showed that the wastewater contained non-biodegradable contents. Hence, based on the Michaelis-Menten equation, a kinetic model considering the non-biodegradable contents in the wastewater was proposed to describe the biodegradation process of COD and NH_3-N in the reactor. With the experimental plots, the kinetic parameters were estimated using 4th-order Runge-Kutta method combined with simplex method, and the simulated curve of the proposed model showed a better fit to the actual response plots than the Michaelis-Menten equation did. It confirmed the rationality of introducing the substrate inhibiting mechanism caused by the non-biodegradable contents in the wastewater.

Key Words: Kinetic model, Operating parameters and Wastewater biodegradation

Introduction

The three-phase biological fluidized bed reactor (BFR), which combines the advantages of microbe immobilization and fluidization technology, has received considerable interests in wastewater treatment due to its high efficient performances [1, 2]. It has been claimed to outperform other bioreactor configurations applied in aerobic wastewater treatment such as activated sludge system and trickling-filters [3, 4].

The fluidization in the reactor is conducted by an upward flow of gas through the bottom of the bed. The bioflim coated carriers along with the wastewater flow upward through the inner tube and downward in the outer annular tube due to density difference between the two regions. Support particles with immobilized biofilms flow along with the liquid. Soluble carbon and nitrogen compounds diffuse through the immobilized biomass and are degraded by microorganisms simultaneously. Microorganisms used for wastewater treatment are usually a mixed culture, the microbial activity of which is closely depended on the composition of the wastewater. Besides, the treatment performance will be affected by the operating conditions. The operating pH and temperature will influence the concerning aerobic biological reaction. The influence of the air supplement represents in its effect on the dissolved oxygen supply and the fluid flow in the reactor.

The theoretical modeling of the three-phase biological fluidized bed reactor has been attempted by several investigators. Some microbial growth kinetic models, either zero or first-order, or Monod equation, were adopted to establish mathematical model for BFR [3]. F. Kargi [5] studied the kinetic of a biological fluidized bed, assumed the reactor as a well mixed homogenous system with no DO (dissolved oxygen) limitations, and supposed that the pollutant degradation reaction followed the Monod equation. However, some kinetic constants determined were not quite in good agreement with literature values. Selene M. A. developed a model to describe the mass transfer process inside a fluidized bed reactor, considering the effects of dispersion, convection and reaction in the liquid phase, as well as diffusion and reaction within the bioflim [6]. The whole methodology related to the transport phenomena in porous media using the method of volume averaging proposed by S. Whitaker, and the numerical solution and the corresponding model parameters were complex. Nevertheless, the standardized kinetic model that could be practically applied to the biological wastewater treatment process in BFR has not yet been established.

The biodegradation of the organic matters includes two basic processes: mass transfer and biochemical reaction. From the view of reaction engineering, the investigation of continuous biological wastewater treatment process in BFR could be boiled down to two parts, one is enzymatic biodegradation reaction and the other is the residential time distribution (RTD) of the reactor. The Michaelis and Menten equation is one of the most widely accepted dynamical reaction models which describe the enzymatic reaction. Mostly, in aerobic wastewater treatment systems contains certain negligible toxic biodegradation organics, the biodegradation kinetic can be best described by introducing a substrate inhibiting mechanism. On the other hand, as non-ideal reactor, the flow pattern of the BFR is considered to be between the CSTR and PFR. However, the RTD of the reactor could be derived by the combination of some ideal and non-ideal flow models. Finally, with the integration of the biodegradation kinetic and the RTD of the reactor, a model could be established to predict the treatment performance of the BFR.

In this paper, a lab-scale BFR was set up with synthetic porous particles made of poly (MA-VAc-MMA) as the biofilm carriers. Systematic experiments were conducted to study the effects of the operating parameters on the treatment performance of the reactor. The study also aimed at developing a kinetic model which described the enzymatic biodegradation reaction in the reactor, considering the substrate inhibiting mechanism. The dynamic behaviors of COD and NH_3-N removals in the reactor were tested, and the corresponding kinetic parameters were determined.

Materials and methods

Three-phase biological fluidized bed reactor

Experiments were conducted in pilot-scale BFR made of organic glass, with a height of 1.15 m and a volume of 8.36 L, shown in Fig. 1. The circulating fluidized section, with a height of 0.66 m, contains a pair of concentric tubes, the inner diameters of which are 105 mm and 70 mm respectively.

The air was provided by a compressor through a distributor at the bottom of the reactor, and the flow rate was controlled by an air rotameter. Wastewater, stored in a reservoir, was pumped into the reactor by a peristaltic pump. Synthetic wastewater composed of glucose, NH_4Cl, K_2HPO_4 with a ratio of COD/N/P=100/5/1 was used through out the experiments. Modest nutrient salts like $MgSO_4$, $CaCl_2$, $NaHCO_3$, etc. were added for a better growth of the microorganisms.

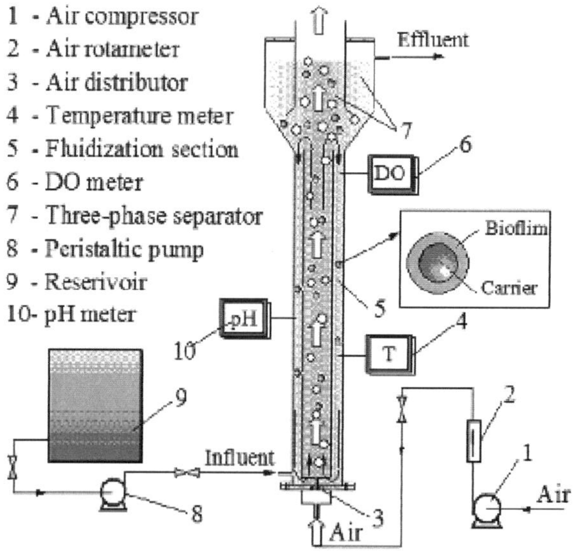

Fig. 1. The experimental set-up

The biomass carriers are the synthetic porous particles made of poly -MA-VAc-MMA with the wet packing density of 1010 kg/m^3. The carriers are almost spherical, and the diameter of the particles ranges from 1 to 2 mm. The inoculum was the activated sludge taken from the biological treatment unit operated at a local municipal wastewater treatment plant. After a short time preculture, the reactor was charged with 8 L activated sludge and 0.36 L carriers. The synthetic wastewater with COD increased from 200 to 400 mg/L was continuously feed into the reactor under the operating HRT of 1.5 h. The inoculation process was performed and continued until the steady operation of the reactor was achieved. The steady operation of the reactor is judged by the stability of biomass concentration on the carriers and the COD removal performance.

Analytical methods

COD was analyzed by fast digestion-spectrophotometric method according to HJ/T 399-2007 of China. Determination of chemical oxygen demand (COD) was carried out using dichromate method, so the COD mentioned in the following refers to COD_{Cr}. NH_3-N was determined using Nessler's reagent colorimetric method according to GB 7479-87 of China. DO and pH were measured by DO analyzer (JBP-607) and pH meter (PHS-3C), respectively. Reproducibility of these measurements was determined by analyzing the same sample three times. Deviations from the average were less than 5% indicating good reliability of measurements.

Modeling and parameters estimation strategies

Kinetic model

The biodegradation kinetic model is based on the assumption that the reactor is a continuously operated fixed film bioreactor system under steady state conditions. The pollutants in the wastewater are the substrate in the biodegradation reaction. The pollutants biodegradation process is a result of the interactions between multiple substrates and enzymes, which could be integrated as an enzymatic reaction.

The Michaelis-Menten equation (1) is one of the classic kinetics model to describe the substrate consumption in the immobilized biologically systems, which can be limited either to first or zero order kinetics based on the substrate concentration.

$$-r_{\text{substrate}} = -\frac{\mathrm{d}C_{\text{substrate}}}{\mathrm{d}t} = \frac{r_{\max} C_{\text{substrate}}}{C_{\text{substrate}} + K_{\mathrm{m}}} \tag{1}$$

Where the $r_{\max}$ is the maximum reaction rate and K_{m} is the Michaelis constant, which is only related to the property of the enzyme.

However, in practical wastewater treatment systems, a substrate inhibiting mechanism should be introduced to the substrate biodegradation kinetics due to the fact that the negligible non-biodegradable organics were inevitable. Thus, equation (1) is modified as follows:

$$-r_{\text{substrate}} = -\frac{\mathrm{d}C_{\text{substrate}}}{\mathrm{d}t} = \frac{r_{\max}(C_{\text{substrate}} - C_{\mathrm{n}})}{C_{\text{substrate}} - C_{\mathrm{n}} + K_{\mathrm{m}}} \tag{2}$$

C_{n} refers to the concentration of the non-biodegradable contents in the wastewater.

Model parameters estimation

Batched experiments were performed to gain the variation of COD and NH_3-N concentration changing with time. With the experiment results, the model parameters were estimated by the followed strategies showed in Fig. 2. The 4th order Runge-Kutta method was applied to solve the differential equations in the model, and the Simplex method was used to adjust the parameters in order that the difference between the experimental and calculated values trend to be a minimum.

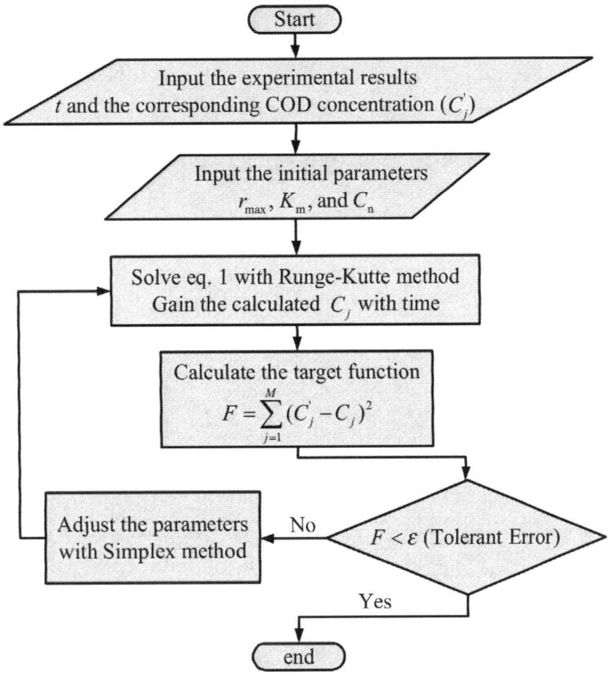

Fig. 2. The strategy of model parameters estimation

Results and discussions

The effects of wastewater contents on treatment performance

In this research, the treatment performance of the reactor was mainly focused on the COD and NH_3-N removals. The variations of the corresponding removal rates with different initial COD and NH_3-N concentrations in the wastewater are depicted in Fig. 3. The COD removal rates increase with the initial COD concentration at low COD levels, while with the further increasing of COD when COD>600 mg/L, the removal rates become nearly stable. The results indicated the inhibition of the biological reaction at lower COD levels, which was probably due to the low biological activity for the lack of carbon source in the substrate for the microorganisms.

Meanwhile, it can be also seen from Fig. 3(b) that the NH_3-N removal rates increase with the initial NH_3-N concentration when NH_3-N is lower than 25 mg/L, and decrease with its further increasing. The activated sludge used in this experiment was taken from a municipal wastewater treatment plant, and its normally adaptive NH_3-N concentration was around 25~35 mg/L. So the overly high or low NH_3-N concentration affected the activity of the nitrobacteria, thus, the inhibition of the nitrification led to the drop of the NH_3-N removal.

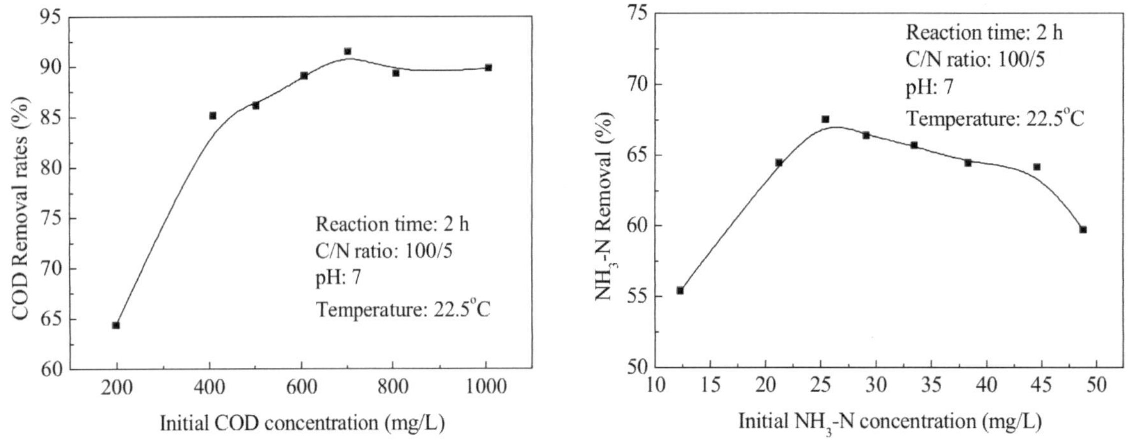

(a) Effect of the initial COD concentration (b) Effect of the initial NH_3-N concentration

Fig. 3. The effect of the initial concentration of the substrate on the treatment performance

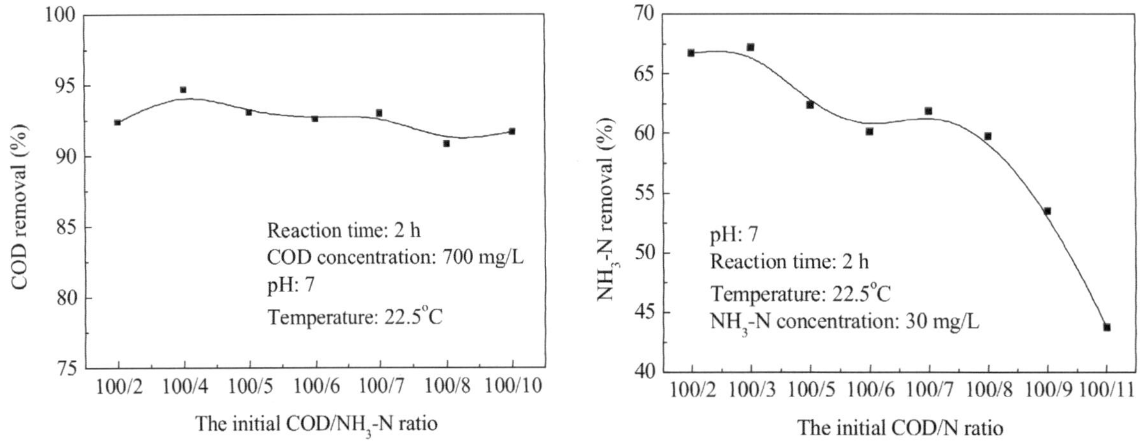

(a) Effect on COD removals (b) Effect on NH_3-N removals

Fig. 4. The effect of the initial COD/NH_3-N on the treatment performance

The effect of COD/NH_3-N ratio on the treatment performance is investigated with initial COD and NH_3-N concentration fixed respectively, and the results are shown in Fig. 4. It can be seen that with the COD/NH_3-N ratio changed from 100/2~100/10, the COD removal rates remains between 90%~95%. The results indicated that effect of COD/NH_3-N ratio on the COD removal was not obvious. While, from Fig. 4(b) we can see that the NH_3-N removal is dramatically affected by the COD/NH_3-N ratio. With the COD/NH_3-N ratio decreases from 100/2 to 100/11, the NH_3-N removal rates drop from 67% to 43%. This may be explained by the fact that the heterotrophic bacteria mainly take charge of the nitrite process in the reactor [7]. With the decrease of COD/NH_3-N ratio, the lack of carbon source led to the inhibition effect on the growth of the nitrifying bacteria, which weakened the nitrification in the reactor.

The effects of operating temperature and pH

The influence of the operating temperature on the treatment performance was investigated through the experiments with fixed wastewater composition and operating pH. As it is shown in Fig. 5, the operating temperature does show certain effects on the treatment performance, especially for the NH_3-N removal. The biodegradation process in the reactor was mainly depended on the enzymatic reaction in the microorganism, and the activities of the enzymes were affected by the surrounding temperature. The experimental results indicated that the modest operating temperature in the experiment was between 22.5~25 ℃, and the nitrifying bacteria was more sensitive to the variation of temperature.

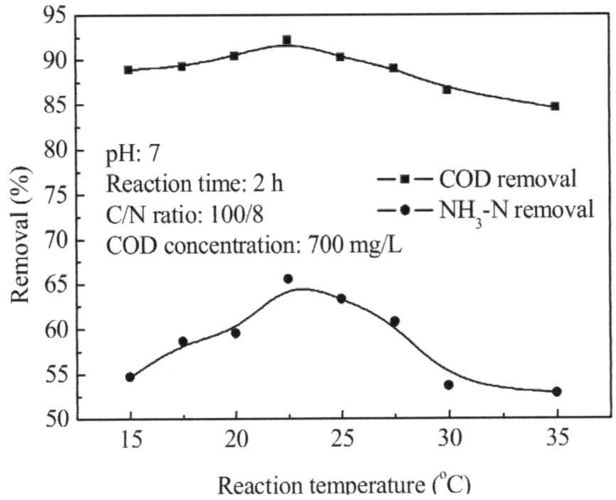

Fig. 5. The effect of operating temperature on the treatment performance

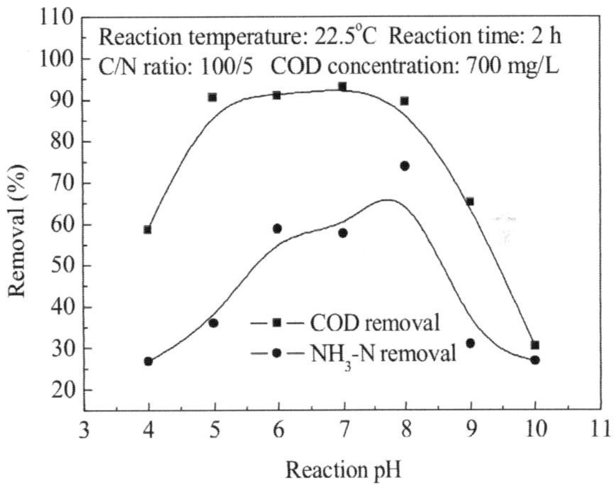

Fig. 6. The effect of operating pH on the treatment performance

The operating pH also exerted a remarkable influence on the biological removals of COD and NH_3-N. The effects of the pH on the treatment performance are shown in Fig. 6. It can be seen that With the pH in the range of 6~8, the COD and NH_3-N removal rates remained on a relatively high level, while beyond the range, the removal rates dropped rapidly. The reason was also attribute to the microorganism activities affected by the surrounding pH.

The effect of air supplement

To investigate the influence of the air flow rate on the treatment performance, the measurements of COD and NH_3-N changing with time were performed in a semi-continuous mode, that is, the air (set at different flow rate) was continuously provided and the wastewater was fed batch-wise at the beginning of each operation. The performance of COD removal in the experiments is shown in Fig. 7(a). It can be seen that the decreasing in COD concentration depends on both time and air flow rate. Under each air supply condition, with the increase of reaction time, the COD concentration decreases and the reduction rate is gradually slowed down. This is due to the decline of microbiological activity caused by the decrease of the substrate concentration. Meanwhile, as can be noted in Fig. 7(a), with the increased air flow rate, the COD removal rate initially increases monotonically and then decreases, attaining the largest removal rate at the air flow rate of 125 L/h. In Fig. 7(b), a similar trend appears in the NH_3-N removal performance. The increased air flow rate had both the positive and negative effects on the treatment efficiency in the reactor.

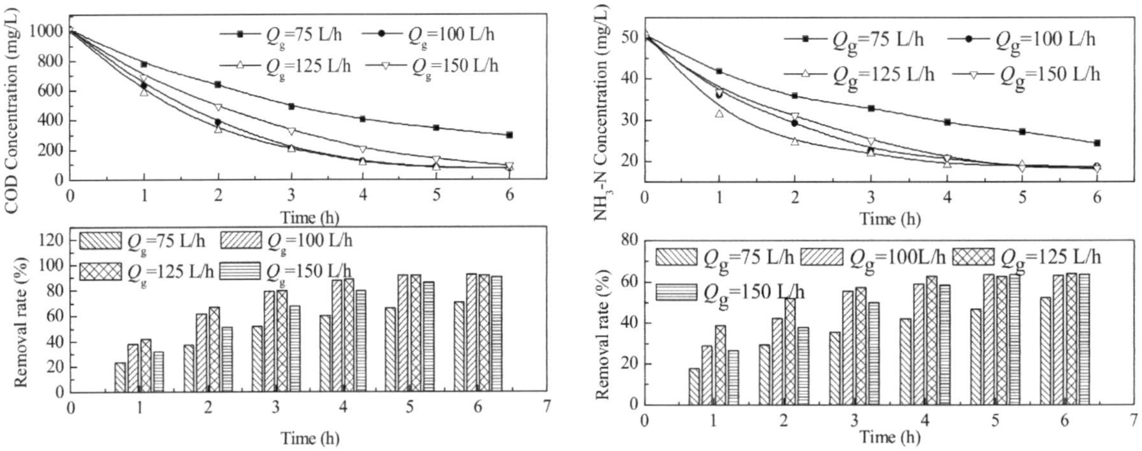

(a) Effect on COD removals (b) Effect on NH_3-N removals

Fig. 7. The effect of the gas flow rate on the treatment performance

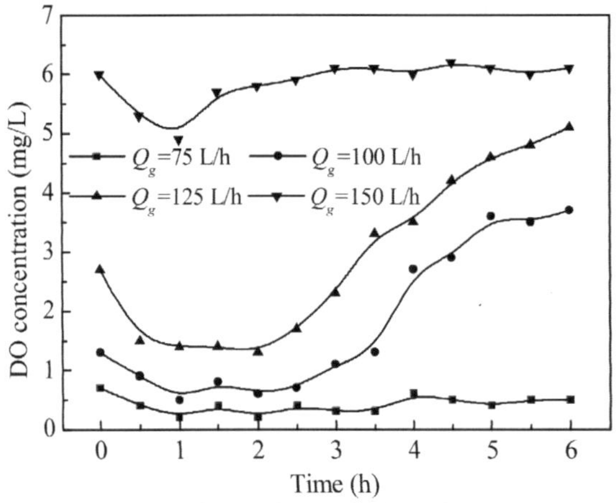

Fig. 8. The variation of DO during the experiments

This can be explained by the fact that with the increase of air flow rate, the oxygen transfer in the reactor was increased, which resulted in the increase of the oxygen supply for the biochemical reaction. The variation of DO during the experiments is shown in Fig. 8. It can be seen that, when the air flow rate is below 125 L/h, anoxic environment is locally appeared in the reactor, for the situation of DO concentration below 1 mg/L occurs in the reactor. Thus, the oxygen becomes the limiting factor in the treatment performance when air flow rate is below 125 L/h. While, with the air flow rate further increased above 125 L/h, the decreased removal efficiency indicating that other factors other than the oxygen supply became the limiting factor. This can be possibly explained that with the increase of air flow rate, the shear force in the fluid increased, resulting in the biomass loss from the carrier surfaces which affects the treatment performance [8, 9]. Consequently, there existed an optimal operating rate, and in this experimental range, 125 L/h was considered as the optimal air flow rate.

Model verification and analysis

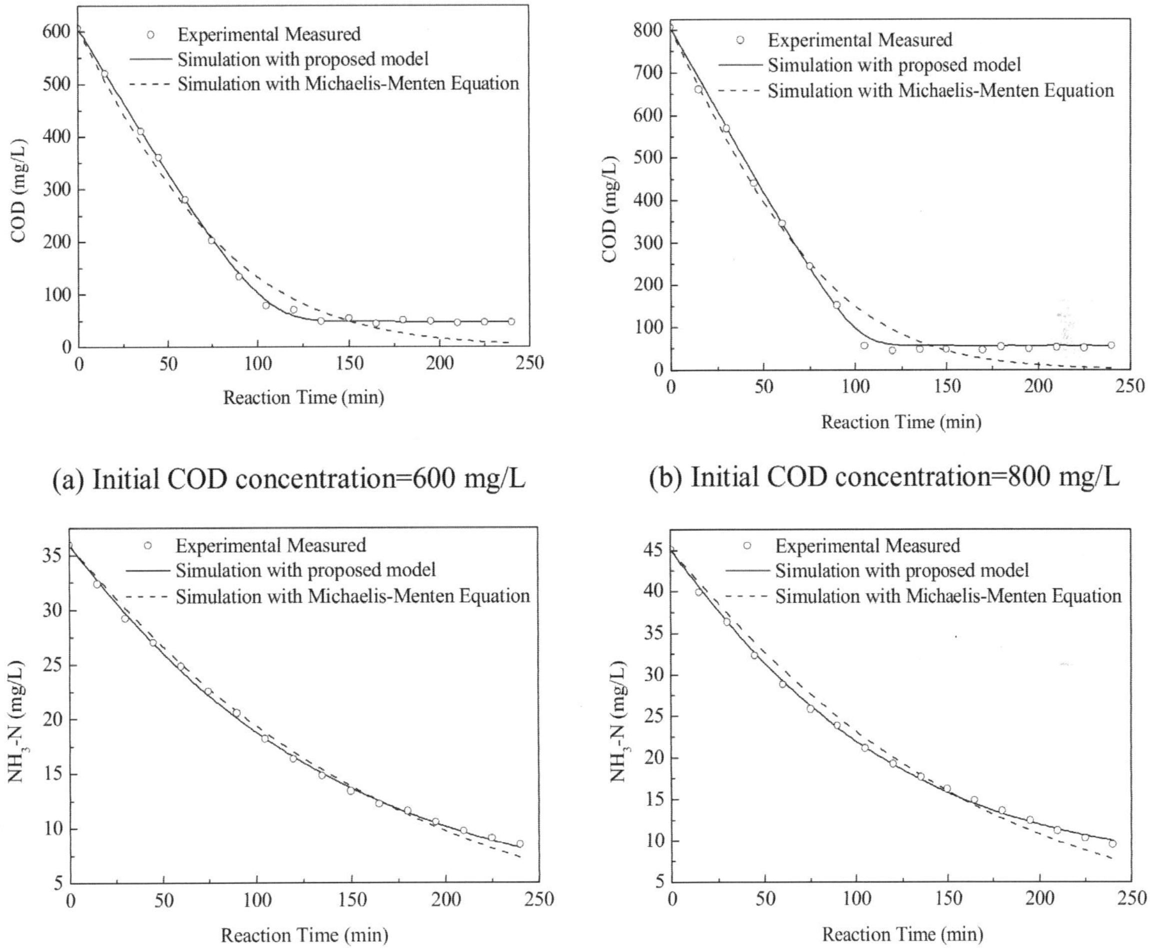

(a) Initial COD concentration=600 mg/L (b) Initial COD concentration=800 mg/L

(c) Initial NH_3-N concentration=35 mg/L (d) Initial NH_3-N concentration=45 mg/L

Fig. 9. Comparison between the experimental results and those simulated by kinetic models

The experimental results obtained from the batched treatment of synthetic wastewater in the BFR are shown in Fig. 9. During the experiments, the initial COD and NH_3-N concentrations were changed, while the COD/NH_3-N ratio was fixed at 100/5. The DO in the reactor was kept around 2.5~3 mg/L to assure no oxygen limitation, and the pH of the wastewater was controlled around 7~8. It can be noted that the COD and NH_3-N decrease with the increasing reaction time, while with the further increase of reaction time, the downtrend becomes smoother. Especially for the COD degradation process, after 120 min of reaction, the COD concentration becomes almost constant as the reaction time increases. It seems like that there are some non-biodegradable contents in the wastewater which restrain the further biodegradation. Thus, as mentioned in 3.1, a kinetic model was proposed based on the Michaelis-Menten equation with the introduction of a substrate inhibiting mechanism, see eq. (2).

With the experimental results, the model parameters for the Michaelis-Menten equation and eq. (2) under each experimental condition were evaluated respectively through the strategy shown in Fig. 2. The evaluated values, including the maximum reaction rate (r_{max}), the Michaelis constant (K_m) and the non-biodegradable contents concentration (C_n), are shown in Table 1. The comparison between the experimental results and simulated results from the models are illustrated in Fig. 9. The simulated curve of the proposed model shows a better fit to the experimental results than the Michaelis-Menten equation does, especially for the reaction after 100 min. The results confirmed the rationality of the introduction of the substrate inhibiting mechanism for the proposed model.

Table 1. The evaluated model parameters

Model parameters	**Initial COD concentration (mg/L)**				**Initial NH_3-N concentration (mg/L)**			
	600	**700**	**800**	**900**	**30**	**35**	**40**	**45**
r_m [mg/(L•min)]	6.17	7.12	8.51	9.71	0.95	0.98	1.02	1.17
K_m (mg/L)	49.64	48.93	47.44	49.77	105.66	106.13	105.51	105.19
C_n (mg/L)	47.70	51.27	57.15	65.63	3.04	3.49	4.39	6.47

At the same time, from Table 1 we can see that the model parameter C_n increases linearly with the increase of the COD and NH_3-N concentration. It can be inferred that some components in the synthetic wastewater caused the inhibition, and the inhibition effect was enhanced by the substrate concentration. The inhibitors were thought to be some minerals and metal ions present in the wastewater, but it remained to be proven. In addition, the maximum reaction rate rises with the substrate concentration, which means in the experimental substrate range, the enzyme activity has not been saturated. The Michaelis constants for COD and NH_3-N are around 49 and 105.5 mg/L respectively, which are in good agreement with the literature values.

Conclusion

The effects of wastewater contents on the treatment performance of a BFR were investigated. The results indicated that COD removal rates increased with the increasing initial COD concentration when it was lower than 600 mg/L, and kept stable with its further increase. The effect of initial NH_3-N concentration on the NH_3-N removal showed a similar trend, and an optimal NH_3-N removal rate was gained when NH_3-N concentration was 25 mg/L. Meanwhile, with fixed COD and NH_3-N concentration, the effects of the COD/NH_3-N ratio on their removal performance were tested respectively. The results showed that the COD/NH_3-N ratio has a negligible effect on the COD removals, while an obvious effect on NH_3-N removals. With the COD/NH_3-N ratio decreased from 100/2 to 100/11, the NH_3-N removal decreased monotonously.

The pH and temperature also showed the effects on the treatment performance. It was found that the favorable range of temperature and pH were 20~25℃ and 6~8, respectively, and beyond which, the treatment performance would be obviously inhibited.

The air flow rate will affect both the fluid flow and the dissolved oxygen supply in the reactor, thus. Within the range of 75~150 L/h, the change of the air flow rate showed both negative and positive effects on the treatment performance, and with the air flow rate of 125 L/h, an optimal treatment was obtained.

The results obtained from the batched treatment experiments in the BFR indicated that the COD and NH_3-N concentration decreased with the increasing reaction time, while the downtrend becomes smoother as the reaction time prolonged further. The further biodegradation was restrained by some non-biodegradable contents in the wastewater. Thus, based on the classic Michaelis-Menten equation, a kinetic model considering the non-biodegradable contents in the wastewater was proposed. The kinetic parameters were estimated using 4th-order Runge-Kutta method combined with simplex method. The theoretical curve of the kinetic model showed a better fit to the experimental resluts than the Michaelis-Menten equation did, which confirmed the rationality of the introduction of the substrate inhibiting mechanism for the proposed model.

References

1. Sokol W., Ambaw A., Woldeyes B. (2009), "Biological wastewater treatment in the inverse fluidised bed reactor," *Chemical Engineering Journal*, 150(1), pp. 63-68.
2. Chan Y. J., Chong M. F., Law C. L., Hassell D. G. (2009), "A review on anaerobic-aerobic treatment of industrial and municipal wastewater," *Chemical Engineering Journal*, 2009, 155(1-2), pp. 1-18.
3. Mowla D., Ahmadi A. (2007), "Theoretical and experimental investigation of biodegradation of hydrocarbon polluted water in a three phase fluidized-bed bioreactor with PVC biofilm support," *Biochemical Engineering Journal*, 36(2), pp. 147-156.
4. Kun H., Changyi J., Yong G., Weixing H. (2011), "Biological Treatment of Wastewater from Sugar Production Process by using Three-Phase Fluidized Bed Reactor," *International Journal of Chemical Reactor Engineering*, September, pp. 1-19.
5. Kargi F., Karapinar I. (1995), "KINETICS OF COD REMOVAL FROM A SYNTHETIC WASTE-WATER IN A CONTINUOUS BIOLOGICAL FLUIDIZED-BED," *Bioprocess Engineering*, 13(5), pp. 265-270.
6. De Souza S., Brandao H. D., de Souza A. A. U. (2008), "Modeling of liquid pollutant biodegradation process in a fluidized bed reactor with biofilm," *Separation and Purification Technology*, 60(2), pp. 162-173.
7. Joo H. S., Hirai M., Shoda M. (2005), "Characteristics of ammonium removal by heterotrophic nitrification-aerobic denitrification by Alcaligenes faecalis no. 4," *Journal of Bioscience and Bioengineering*, 100(2), pp. 184-191.
8. Tijhuis L., Hijman B., Van Loosdrecht M. C. M., Heijnen J. J. (1996), "Influence of detachment, substrate loading and reactor scale on the formation of biofilms in airlift reactors," *Applied microbiology*, 45(1-2), pp. 7-17.
9. Cao Y. S., Alaerts G. J. (1995), "INFLUENCE OF REACTOR TYPE AND SHEAR-STRESS ON AEROBIC BIOFILM MORPHOLOGY, POPULATION AND KINETICS," *Water Research*, 29(1), pp. 107-118.

(684a) Bench-Scale Testing of Coal Conversion Using Moving Bed in Coal Direct Chemical Looping Process

Siwei Luo[1], Omar McGiveron[2], Liang Zeng[1], Samuel Bayham[3] and L. - S. Fan[1], (1)William G. Lowrie Department of Chemical and Biomolecular Engineering, The Ohio State University, Columbus, OH, (2)William G. Lowrie Department of Chemical and Biomolecular Engineering, Ohio State University, Columbus, OH, (3)William G. Lowrie Department of Chemical and Biomolecular Engineering , Columbus, OH

In comparison with gaseous fuel looping processes, the coal direct chemical looping (CDCL) process directly converts coal in the reducer, eliminating the need for the gasifier and air separation unit (ASU). A bench scale moving bed reactor was used to study the coal conversion. In a typical test, a mixture of oxygen carrier particles and solid fuel was introduced via a Y-shaped pipe. The particle was then steadily moved downward through the heated zone where it reacts with the coal powder. Meanwhile, the enhancing gas CO_2 was introduced into the system together with carrier gas N_2. During the experiment, the bottom screw continuously conveyed the solids from the reactor into a funnel below the screw feeder at a pre-calibrated rotational speed. Almost complete coal conversion could be achieved using the composite iron oxides particles. Various post-experiments have also been conducted on iron oxides particles. The operational results prove that a moving bed reactor design is favorable for CDCL application.

(684b) Chemical Looping – A Novel Process for Partial Oxidation of Methane

Saurabh Bhavsar[1,2], Kenneth Nadeau[2] and Götz Veser[1,2], (1)US DOE-National Energy Technology Laboratory, Pittsburgh, PA, (2)Department of Chemical and Petroleum Engineering, University of Pittsburgh, Pittsburgh, PA

Chemical looping combustion (CLC) is an emerging clean combustion technology which offers an efficient and elegant route for fossil fuel combustion with inherent CO_2 capture. In CLC, an oxygen carrier (typically a metal) undergoes cyclic reduction and oxidation with fuel and air, respectively. After condensation of the steam from effluent of the reducer reactor, a high pressure, high purity sequestration ready CO_2 stream is obtained. Thus CLC provides efficient, NO_x-lean, flameless combustion route with inherent CO_2capture.

While the efficiency of chemical looping as a combustion technology has been demonstrated for a broad range of fuels to-date, the chemical looping principle can also be extended onto other reactions, including reforming and partial oxidation reactions. We have previously demonstrated that chemical looping reforming offers an attractive process alternative to conventional reforming by replacing air with steam and/or CO_2 as oxidant. This allows for novel and efficient ways to produce hydrogen and activate CO_2via reduction to CO, respectively.

In the present contribution, we demonstrate that chemical looping is also an attractive process variant for partial oxidation reactions. By controlling the oxidation state of the oxygen carrier, total oxidation of the fuel can be avoided, and chemical looping hence opens a novel mode of reactor operation for highly efficient partial oxidation. This principle is demonstrated here with methane as fuel, resulting in catalytic partial oxidation of methane to synthesis gas (CPOM). The attractiveness of this novel CPOM process lies in the fact that it is possible to operate the process with air (rather than pure oxygen) without diluting the resulting syngas streams.

In our experiments, Fe and Ni supported on CeO_2 were chosen as oxygen carriers for chemical looping CPOM in a fixed bed reactor. Ni-CeO_2 and Fe-CeO_2 were synthesized by simple incipient wetness, deposition precipitation technique and characterized by TEM, SEM, XRD, BET, and TPR/TPO, and then evaluated in fixed bed reactor studies as a function of oxidation state. Our results show that by limiting the duration of air oxidation of the carrier, CH_4can be selectively converted to syngas with high conversion. Interestingly, the use of a reducible support is critical for the process: Comparative studies with alumina-supported carriers show reduced performance, both in terms of syngas selectivity and methane conversion.

Overall, our results demonstrate that chemical looping offers a novel, highly attractive process variant for catalytic partial oxidation of methane, and thus suggest that the chemical looping principle could find application for a broader range of partial and selective oxidation reactions.

(684c) Reaction Rate Analysis Relevant to Chemical Looping with Oxygen Uncoupling for Solid Fuels

Asad H. Sahir[1], JoAnn S. Lighty[1], Hong Yong Sohn[2] and Henrik Leion[3], (1)Chemical Engineering, University of Utah, Salt Lake City, UT, (2)Department of Metallurgical Engineering, University of Utah, Salt Lake City, UT, (3)Department of Chemical & Biological Engineering, Chalmers University of Technology, Gothenburg, Sweden

Technologies based on the chemical-looping combustion concept have invited significant research attention, because of their potential in obtaining a high purity CO_2 stream with reduced energy penalty during the fuel combustion process. Chemical-looping with oxygen uncoupling (CLOU) involves two interconnected fluidized bed reactors, an air and a fuel reactor. The combustion of fuel takes place in the presence of oxygen that is released by the decomposition of a metal oxide (e.g. CuO) which serves as an oxygen carrier in a fuel reactor. The reduced metal oxide (Cu_2O) is subsequently re-oxidized in the presence of atmospheric oxygen in a separate air reactor.

In this talk, a methodology for calculating reaction rates based on the data obtained from previously published batch fluidized bed experiments on CLOU will be discussed. These experiments were done by alternating between air and fuel addition in the same reactor. During fuel cycles, the role of thermodynamics in conjunction with chemical kinetics for analyzing CuO decomposition reaction to release oxygen will be highlighted. A procedure to elucidate the solid fuel oxidation kinetics from batch-scale fluidized bed CLOU experiments will also be discussed. The Law of Additive Reaction Times has been used to analyze the Cu_2O oxidation reaction for the air reactor in this study. The utility of the rate analysis to offer insights for developing process engineering simulations for CLOU will also be discussed.

(684d) Holistic Kinetic Study of the Reduction of CH4 with NiO in Chemical-Looping Combustion

Zhiquan Zhou and George M. Bollas, Department of Chemical, Materials & Biomolecular Engineering, University of Connecticut, Storrs, CT

The objective of this work is to compare reaction schemes and kinetic mechanisms that accurately describe the dynamic behavior in chemical-looping reducers, operating with CH_4 and NiO/support (oxygen carrier).

The reaction scheme proposed is composed of catalytic gaseous reactions and heterogeneous reactions with the oxygen carrier include reactions of CH_4, CO and H_2. Ni-catalyzed gaseous reactions considered in this scheme include steam reforming, overall steam reforming, water gas shift, dry reforming, methane decomposition and carbon gasification by CO_2 and H_2O.

Kinetic expressions reported in the literature are compared and their parameters are estimated on the basis of experimental data of Ni-based oxygen carriers for chemical-looping combustion of CH_4. Rates of heterogeneous reactions based on shrinking core model and modified volumetric model with reported data from literature are compared.

Dynamic parameter estimation is performed to estimate kinetics of Ni-based oxygen carriers for chemical-looping reduction. Pre-exponential factors and activation energies for the various mechanisms reported reactions of the scheme proposed are fitted to dynamic data of chemical-looping reduction of NiO using CH_4. CLC data from the literature vary in terms of reactor volume, space velocity, and temperature, generating a sufficient span of conditions to explore the kinetic mechanisms most accurate for CLC. The consistency and discrepancy between the best fit kinetic parameters of different units will be presented.

The capability of the model to predict the gas composition at various operating conditions and reactor configurations using the same reaction mechanisms will be shown, allowing for confidence in the kinetics used in other types of reactor models (e.g., bubbling beds, risers, circulating fluidized beds) and a comprehensive analysis of the scale-up capabilities of CLC reducers.

Acknowledgement: This material is based upon work supported by the National Science Foundation under Grant No. 1054718.

(684e) Modeling Chemical-Looping Combustion in Bubbling Fluidized Bed Reactors

Zhiquan Zhou, Monica Navarro and George M. Bollas, Department of Chemical, Materials & Biomolecular Engineering, University of Connecticut, Storrs, CT

Chemical-looping combustion (CLC) is typically realized as a dual fluidized bed technology. Most current CLC reactor designs for gaseous fuels processing comprise riser and bubbling fluidized beds [1], two bubbling fluidized beds [2], or dual circulating fluidized beds [3]. Because of the higher space times required for complete metal oxide reduction (or fuel oxidation), when compared with the metal oxidation reactions, fuel reactor designs have focused on bubbling bed technologies. The objective of this work is to analyze existing bubbling bed reactors for chemical-looping reduction through dynamic modeling. Experimental data from published bubbling bed chemical-looping reactors are used for model development and validation. The reaction scheme and kinetics used in this study have been developed and validated against published fixed bed chemical-looping units operating with CH_4 and NiO/support (oxygen carrier) [4]. The proposed reaction scheme includes heterogeneous reactions of NiO reduction by CH_4, H_2 and CO, gas phase reactions including reforming reactions (steam, dry and overall steam), water gas shift reaction, methane decomposition and carbon gasification by CO_2 and H_2O. The two-phase (emulsion and bubble) flow model is used to predict bubbling bed hydrodynamics [5,6]. It considers gas bubbles flowing through a dense emulsion phase at a relative velocity given by minimum fluidization with gas percolating through the bed of solids (shown in Figure 1). Using this model the gas profiles predicted for the data published by Iliuta et al. [7] are shown in Figure 2. Gas composition profiles are in good agreement with experiment data.

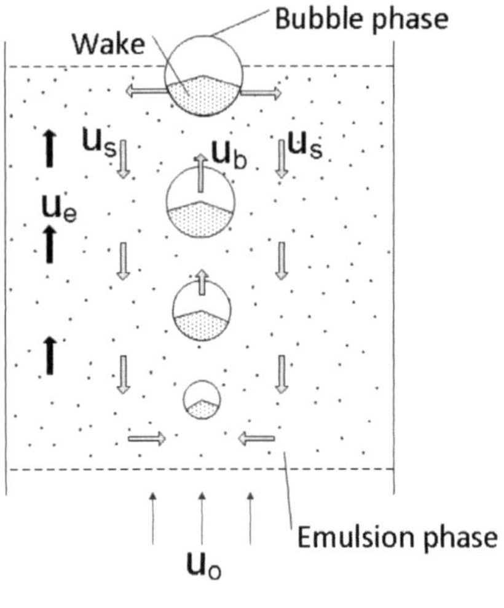

Figure 1: *Bubbling fluidized bed reactor physical model*

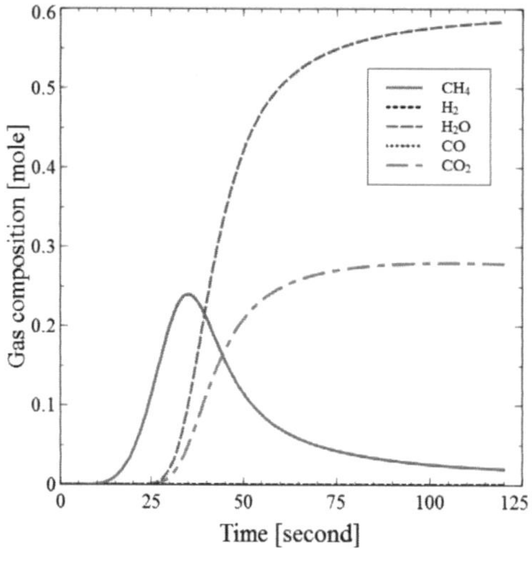

Figure 2: *Chemical-looping selectivity using Ni/NiO and CH_4 in bubbling fluidized bed at 645°C*

A summary of the bubbling bed experimental data analyzed and modeled in this work, including all the relevant published data for CLC with CH_4 and NiO as the oxygen carrier, is presented in Table 1.

Table 1: *CLC bubbling bed units*

Author	Iliuta et al. [7]	Hoteit et al. [8]	Jerndal et al. [9]	Gayán et al. [10]	Ryu et al. [11]
T(°C)	623 – 810	800 – 900	950	950	900
NiO/Support	15% NiO/Al_2O_3	60% $NiO/NiAl_2O_4$	40% $NiO/\alpha Al_2O_3$	18% NiO/ αAl_2O_3 28% NiO/ γAl_2O_3	60% NiO/bentonite 70% $NiO/NiAl_2O_4$
OC load (kg)	0.3	2.5 – 4	0.015	0.3	2.3
Particle size (μm)	140	171	125 – 180	100 – 300	106 – 212
Specific surface area (m^2/g)	102	7	0.91	7, 77.5	-
CH_4 composition	10%, 50% in Ar	100%	100%	25% in N_2	45% syngas in N_2 H_2: CO_2: CO = 30:10:60
Gas flow rate (m^3/s)	2E-05	2E-05 – 2E-04	7.5E-06	2E-04	3.33E-05
I.D. (mm)	46	96	22	54	50
Bed height(m)	0.23	1	0.0179	~0.1	0.4
Space time (s $gNiO^0$ /gCH_4)	5297	9747	1254	1643	-
Bulk density (kg/m^3)	785	2200	2400	2500, 1800	1172

In this presentation, implementation of this model in different bubbling fluidized bed units of the various CLC laboratories will be shown. A comparison of the predicted results with the experimental results for the different reactors will be illustrated and discussed. Dynamic parameter estimation is performed to re-optimize the kinetics of Ni-based oxygen carriers for chemical-looping reduction in bubbling fluidized beds. Best fit kinetic parameters between different bubbling bed units will be presented, along with the consistency and discrepancy of the kinetic parameters estimated for their fixed bed reactor equivalents.

Acknowledgement: This material is based upon work supported by the National Science Foundation under Grant No. 1054718.

References:

[1] A. Lyngfelt, B. Leckner, T. Mattisson, Chemical Engineering Science 56 (2001) 3101-3113.

[2] G. Adanez Juan Francisco, L.F.D. Diego, P. Gaya, J. Celaya, A. Abad, Ind. Eng. Chem. Res. (2006) 2617-2625.

[3] T. Proll, P. Kolbitsch, J. Bolhar-Nordenkampf, H. Hofbauer, AIChE Journal 55 (2009) 3255-3266.

[4] Z. Zhou, G. Bollas, AIChE Annual Meeting, Pittsburgh, PA (2012).

[5] D. Kunii, O. Levenspiel, I & EC Fundamentals 7 (1968) 446-452.

[6] D. Kunii, O. Levenspiel, I & EC Process Design and Development 7 (1968) 481-492.

[7] I. Iliuta, R. Tahoces, G.S. Patience, S. Rifflart, F. Luck, AIChE Journal 56 (2010) 1063-1079.

[8] A. Hoteit, M.K. Chandel, A. Delebarre, Chemical Engineering & Technology 32 (2009) 443-449.

[9] E. Jerndal, T. Mattisson, A. Lyngfelt, Energy 94 (2009) 665-676.

[10] P. Gayán, C. Dueso, A. Abad, J. Adanez, L.F. de Diego, F. García-Labiano, Fuel 88 (2009) 1016-1023.

[11] H.-J. Ryu, D. Shun, D.-H. Bae, M.-H. Park, Korean Journal of Chemical Engineering 26 (2009) 523-527.

(684f) Dynamic 2D Heterogeneous Model for Chemical-Looping Combustion in Fixed Bed Reactors

Lu Han, Zhiquan Zhou and George M. Bollas, Department of Chemical, Materials & Biomolecular Engineering, University of Connecticut, Storrs, CT

Dynamic 2D heterogeneous model for chemical-looping combustion in fixed bed reactors

Lu Han, Zhiquan Zhou, and George M. Bollas, Department of Chemical, Materials & Biomolecular Engineering, University of Connecticut, Storrs, CT

Abstract:

The objective of this work is to present a dynamic, heterogeneous two-dimensional model for fixed bed chemical-looping reactors. This model can be applied to chemical-looping reforming processes and is suitable for use with large catalyst pellets in industrial-scale applications as well as small catalyst pellets in small-scale units. This model is applied to modeling the reduction step in chemical-looping combustion to examine the dynamic behavior and process sensitivity in the fixed bed reactors.

The reactions involved in the chemical-looping reduction reaction step are:

Non-catalytic gas-solid reactions:	
$CH_4 + 2NiO \leftrightarrow 2Ni + 2H_2 + CO_2$ $H_2 + NiO \leftrightarrow Ni + H_2O$ $CO + NiO \leftrightarrow Ni + CO_2$ $CH_4 + NiO \leftrightarrow Ni + 2H_2 + CO$	
Catalytic gas-solid reactions:	
$CH_4 + H_2O \overset{Ni}{\leftrightarrow} CO + 3H_2$ $CH_4 + CO_2 \overset{Ni}{\leftrightarrow} 2CO + 2H_2$ $CH_4 + Ni \overset{Ni}{\leftrightarrow} Ni - C + 2H_2$	$CO + 3H_2 \overset{Ni}{\leftrightarrow} CH_4 + H_2O$ $CO + H_2O \overset{Ni}{\leftrightarrow} CO_2 + H_2$ $C + CO_2 \overset{Ni}{\leftrightarrow} 2CO$ $C + H_2O \overset{Ni}{\leftrightarrow} CO + H_2$

The kinetics for all reactions used follows the results by Zhou, et al. who developed a kinetics network capable of modeling various published experimental data.

A mathematical model is developed with the assumptions of an axially dispersed plug flow with negligible mass or heat variations in the radial direction of the reactor, external and internal transport effects to the catalyst particles, and reaction occurring inside the pores of the oxygen carrier. The model equations used (shown below) consider the mass, energy, and momentum balances of the gas along the reactor, and the mass and energy balances inside the catalyst pellet. The gas phase momentum balance is assumed to be in pseudo-steady state in which the Ergun equation is applied.

Bulk phase	Catalyst phase
$\varepsilon_b \frac{\partial C_i}{\partial t} + \frac{1}{A_c}\frac{\partial F_i}{\partial z} = \varepsilon_b D_{ax,j}\left(\frac{\partial^2 C_i}{\partial z^2}\right) + k_{c,i} a_v \left(C_{s,i}\vert_{R_p} - C_i\right)$	$\varepsilon_c \frac{\partial C_{s,i}}{\partial t} = \frac{1}{r^2}\frac{\partial}{\partial r}\left(D_{eff,i} r^2 \frac{\partial C_{s,i}}{\partial r}\right) + \rho_s \sum R_i$

$$\varepsilon_b C_{p,f} C_T \frac{\partial T}{\partial t} + \frac{C_{p,f} F_T}{A_c}\frac{\partial T}{\partial z} = \varepsilon_b \lambda_g \left(\frac{\partial^2 T}{\partial z^2}\right) + h_f a_v \left(T_p\vert_{R_p} - T\right)$$

$$\left((1-\varepsilon_c)\,\rho_s C_{p,s} + \varepsilon_c C_{p,cat} C_{T,cat}\right)\frac{\partial T_s}{\partial t} = \frac{\lambda_s}{r^2}\frac{\partial}{\partial r}\left(r^2 \frac{\partial T_s}{\partial r}\right) + \rho_s \sum (-\Delta H_i)(-R_i)$$

In this presentation, model predictions will be compared against experimental data from various laboratories utilizing fixed bed reactors for chemical-looping combustion and reforming. Process sensitivities to pressure changes, catalyst characteristics, thermal gradients, and diffusional limitations will be illustrated and discussed.

Acknowledgement: This material is based upon work supported by the National Science Foundation under Grant No. 1054718.

(684g) Three-Dimensional Reacting Flow Simulation of a Chemical Looping Fuel Reactor

Yong Liu, National Energy Technology Laboratory, U.S. Department of Energy, Pittsburgh, PA, E. David Huckaby, Computational Science Division, National Energy technology Laboratory, Morgantown, WV, Michael Gallagher, URS Corporation, Morgantown, WV and Stephen Carpenter, NETL-URS, Morgantown, WV

Chemical looping combustion (CLC) is one of several proposed methods for energy production from fossil fuels with CO_2 capture. The most common CLC configuration consists of an interconnected fast fluidized bed air reactor and a bubbling bed fuel reactor. The fuel reactor is where the fuel, such as CH_4, H_2 or even coal, reacts with the solid oxygen carrier. Thus the performance of the fuel reactor will determine the performance of the whole interconnected system. Simulations using ANSYS-Fluent were performed and combined with user defined functions to study the flow behavior and the reduction of the oxygen carrier in the fuel reactor. Parameters such as the oxygen carrier particle size, bed height in the fuel reactor, and fluidization gas flow rates were varied to study the effects on the performance of the fuel reactor. The distributions of pressure, temperature and gaseous components such as N_2, CO_2, H_2O and CH_4 were analyzed and the results will be discussed.

(739a) Clc Coal Combustion in a 10 Kw Pilot Plant

Tiago Sozinho[1], William Pelletant[1], Florent Guillou[1], Thierry Gauthier[1] and Hélène Stainton[2], (1)IFP Energies Nouvelles, Solaize, France, (2)Total Gaz Energies Nouvelles, Paris, France

Chemical looping combustion (CLC) is a technology in which the oxygen required for the fuel combustion reaction is supplied by a metal oxide (oxygen carriers, OC). The principle is based on the continuous circulation of OC particles between two interconnected reactors. In a first reactor, called "Air reactor", the oxygen carrier is oxidized by air . Then the OC particles are conveyed to a second reactor, called "Fuel reactor", where they are reduced by a fuel stream. Finally, the oxygen carrier particles are recycled to the Air reactor to start a new cycle. Therefore, direct contact between air and fuel is prevented, leading to the formation of a rich CO_2-H_2O flue gas, which can be easily condensed and captured.

Contrariwise to post-combustion or oxycombustion technologies, in chemical looping combustion the CO_2 separation is inherent to the process and therefore no energy is expended for the capture [1]. Thanks to this, chemical looping combustion is considered as one of the most promising CO_2 capture technology.

As part of a partnership between IFPEN and TOTAL, a 10kWth pilot plant has been built in 2010 at IFPEN facilities. This CLC prototype unit is composed of three interconnected bubbling fluidized bed reactors, which are separated by loop seals that prevent gas leakage between reaction zones. The pilot solid circulation rate is controlled by non mechanical L-valves adapted to high temperature operation.

After a first period of tests where the pilot was validated on gaseous feeds [2], the prototype was modified for solid fuels testing in the beginning of 2011. An additional capacity was added to separate the unburned carbon from the oxygen carrier, and a solid fuel dosing injection system was implemented on the fuel reactor. The paper will present some important results obtained from the continuous 10kWth pilot operation with South African coal. Among the results, impact of different important parameters affecting coal conversion will be exposed, such as : - Temperature effect between 850°C and 940°C, - Residence time between 5 and 30min, - Steam partial pressure.

References:

[1] Figueroa *et al.*, "Advances in CO_2 capture technology – The US DOE carbon sequestration program", Int. J. of greenhouse gas control, 2008, vol. 2 9-20

[2] Hoteit *et al.* ECCRIA 2010

(739b) Sub-Pilot Scale Demonstration of Iron-Based Coal-Direct Chemical Looping Combustion Process

Hyung Kim, Chemcial biomolecular Engineering, Ohio State University, Columbus, OH and L. - S. Fan, William G. Lowrie Department of Chemical and Biomolecular Engineering, The Ohio State University, Columbus, OH

With the fast growing demand of power generation, excessive CO_2 emission from the coal-fired power plants becomes the environmental concern that causes the global warming. In order to regulate the CO_2 emission, many carbon capture technologies have been developed to retrofit to the coal-fired power plants. Among these newly developed technologies, the chemical looping processes exhibit the promising economic and technical feasibility to capture CO_2 from the coal conversion processes. At the Ohio State University, the Coal-Direct Chemical Looping (CDCL) combustion process using iron-based oxygen carriers was developed as coal conversion process with in-situ CO_2 capture. In the previous studies of the CDCL process, the oxygen carrier particles were synthesized and the process scheme with a packed moving bed was proposed. The CDCL system mainly consists of three components; reducer, combustor and riser. In the reducer, reduction of Fe_2O_3 particles yields a mixture of Fe and FeO particles, while converting the coal into CO_2 and steam, which can be readily available for sequestration. The combustor is a fluidized bed that oxidizes the reduced particles back to Fe_2O_3 with air. The oxidation of iron is an exothermic reaction and the heat can be recovered to be used in the coal power plant system. In the riser, there is no chemical reaction, but it pneumatically transports the particles to the reducer.

Objective of this project is to demonstrate the continuous operation of $25kW_{th}$ CDCL sub-pilot unit with inputs of various solid fuels. Before the sub-pilot unit is designed and constructed, the cold flow model (CFM) with near 1:1 scale to sub-pilot unit was constructed to have better understanding of the reactor models and develop the solid handling strategy in the actual CDCL sub-pilot unit. From the experience of CFM study, the final design of sub-pilot unit was obtained and the reactor parts are fabricated for the assembly. Moreover, all the peripheral equipments for the sub-pilot unit such as coal feeder, gas injection and sampling system and analysis systems were prepared. The sub-pilot unit was successfully operated with metallurgical coke and sub-bituminous coal. With capability to handle hot solids in the sub-pilot unit, the demonstration works can move forward for more parametric studies, which will be used for the techno-economic studies.

(739c) Iron Based Chemical Looping Gasification Using Gaseous Fuels

Liang Zeng, Andrew Tong, Hyung Rae Kim, Dawei Wang, Mandar Kathe, Siwei Luo, Qiang Zhou, Elena Chung, Zhenchao Sun and Liang-Shih Fan, William G. Lowrie Department of Chemical and Biomolecular Engineering, The Ohio State University, Columbus, OH

The chemical looping gasification (CLG) process circulates iron-based oxygen carrier particles through 3 reactors, namely reducer, oxidizer, and combustor, to perform reduction-oxidation reaction cycles for carbon dioxide capture and hydrogen production from various carbonaceous fuels such as natural gas and coal derived syngas. The countercurrent moving-bed operation of the reducer highlights the Ohio State University CLG technology. It ensures the full conversion of fuels while enhancing the extent of iron based oxygen carrier conversion. The high extent of iron based oxygen carrier conversion facilitates the generation of hydrogen from steam-iron reaction in the oxidizer. The combustor fully regenerates the oxygen carrier particles, which are then circulated back to the reducer via a riser. The CLG process can be configured to generate hydrogen, electricity or any combination thereof. It can also be incorporated into the Fischer-Tropsch process to enhance the efficiency and yield of coal to liquid fuels production. The process has been tested under a 25kW_{th} sub-pilot scale continuous operation for more than 300 hours with current efforts focused on the scale up demonstration of the 250 kW_{th} pilot CLG unit.

The design and operation of the 25 kW_{th} CLG sub-pilot unit will be the focus of this presentation. The design philosophy of the sub-pilot unit will be elucidated initially, with key emphasis on influence of the mode of operation. Multiple parametric operation studies for various gaseous fuels and system performance with regards to fuel conversions and hydrogen production will also be presented. Both multistage equilibrium modeling and dynamic reactive flow modeling have been conducted to compare the experimental results with the theoretical thermodynamic and kinetic conversions. Finally, the scale-up design and construction of the 250 kW_{th} pilot-scale unit will be updated.

(739d) Exergetic Analysis of Chemical Looping Reforming Processes

Michelle Najera, Department of Chemical Engineering, University of Pittsburgh, Mascaro Center for Sustainable Innovation, University of Pittsburgh, Pittsburgh, PA, Saurabh Bhavsar, Department of Chemical Engineering, University of Pittsburgh, US DOE-National Energy Technology Laboratory, Pittsburgh, Pittsburgh, PA, Neha Nandaku, Department of Chemical Engineering, Carnegie Mellon University and Götz Veser, Department of Chemical and Petroleum Engineering, University of Pittsburgh, Pittsburgh, PA

Chemical looping combustion (CLC) is an emerging clean combustion technology in which metal oxygen carriers are exposed to oxidation/reduction cycles with air and fuel, respectively. The oxygen carrier is first oxidized by air in the oxidizer, forming a metal oxide, and then reduced by a fuel in the reducer, forming CO_2, water, and the reduced form of the metal. Overall, CLC yields a flameless, NOx-free combustion and produces sequestration-ready CO_2. CLC has long been recognized as a highly efficient energy process when compared to conventional combustion. This increased energy efficiency is achieved via the division of the conventional gas phase combustion reaction into the two reversible gas-solid reactions in the CLC oxidizer and reducer. The high efficiency is due to a reduction in the entropic losses, or irreversibilities, associated with the combustion process, i.e. chemical looping has high exergetic efficiency, resulting in more usable energy when compared to conventional gas phase combustion.

Beyond CLC, several chemical looping reforming (CLR) technologies have also been identified. For instance, the use of steam instead of air as the oxidant has been demonstrated by our group and others to yield an efficient route to produce high-purity H_2 as the oxidizer effluent. Similarly, our group has recently proposed the use of CO_2 instead of air as the oxidant, resulting in chemical looping dry reforming (CLDR) as an novel method to utilize CO_2 via reduction to CO (i.e. via "CO_2 activation"). While the thermodynamic and kinetic feasibility was demonstrated for these chemical looping reforming processes, their exergetic efficiency has not been studied to-date.

In the present work, we have developed a process model for CLR processes and implemented it in the Aspen modeling software. First results from these studies indicate an increased exergetic efficiency of chemical looping dry reforming (CLDR) compared to conventional dry reforming. Extension onto H_2 production via CLR processes and comparison to conventional H_2 production via steam reforming and partial oxidation is currently under way.

(739e) Chemical Looping Combustion (CLC) for CO2 Capture in Coal-Based Integrated Gasification Combined Cycle (IGCC) Power Plants – Performance and Cost Analysis

Hari C. Mantripragada and Edward S. Rubin, Engineering and Public Policy, Carnegie Mellon University, Pittsburgh, PA

Chemical looping combustion (CLC) is an indirect combustion process in which fuel is combusted without direct contact with air. Transfer of oxygen between air and fuel takes place with the aid of an oxygen-carrier (OC), usually oxides of transient metals. The reduced form of metal oxide extracts O_2 from air in one reactor and then transfers it to fuel in a subsequent reactor. Since the fuel does not come in direct contact with air, the products of combustion contain only carbon dioxide (CO_2) and water (H_2O). CO_2 stream of very high purity can be obtained by condensing the water vapor. Gaseous fuels such as natural gas or syngas (CO and H_2) can be used as fuels in CLC, for inherent CO_2 capture. The focus of this paper is the use of CLC for CO_2 capture in a coal-based integrated combined cycle power plant (IGCC) using two different configurations (high pressure and atmospheric pressure), in terms of the overall thermodynamics and cost.

CO_2 capture in conventional IGCC designs involves a water gas shift reaction to convert CO in the syngas from gasifier to CO_2 and subsequent capture of CO_2 using a physical absorption solvent (Selexol or Rectisol). Alternatively, CLC can be used for CO_2 capture in a coal-based IGCC. The products from the CLC reactor system are two separate streams of oxygen-depleted air and combustion products (CO_2 and H_2O), both at high temperature. Two configurations of IGCC power plants using CLC can be formulated – high pressure CLC and atmospheric pressure CLC. In the high pressure configuration, CLC process takes place at a high pressure and the exhaust streams from both reactors can be expanded in separate gas turbines to generate electricity. The hot exhaust from these turbines can be used to generate steam in a heat recovery steam generator (HRSG) for further electricity production. Water can be condensed from the CO_2-stream in order to obtain high purity CO_2 for further compression and storage. On the other hand, in the atmospheric pressure CLC configuration, CLC process occurs at atmospheric pressure and the hot exhaust streams from the reactors are cooled down to generate steam in HRSG, which is used to generate electricity in steam turbines. The advantage of atmospheric pressure CLC is the absence of air compressor which is required in the high-pressure CLC configuration. However, the reactor size increases with decreasing pressure which will affect the capital cost of the system. The relative advantages and disadvantages of both these configurations will be studied in terms of their thermodynamic performance as well as capital and operating costs.

Performance and cost models are developed to perform a detailed thermodynamic and economic analysis of an IGCC plant using CLC for CO_2 capture. Chemical equilibrium models will be used to simulate the CLC reactions, using NiO/Ni CuO/Cu and Fe_2O_3/Fe_3O_4 as the oxygen carriers. The effect of varying syngas compositions and operating conditions such as temperature will also be studied. A comparison will be made with a conventional IGCC power plant with CO_2 capture in order to understand the relative feasibility of CLC when applied to IGCC.

(739f) Single Step NO, SO2 and CO2 Removal From Gas Mixtures Using Calcium Sorbent and Char - Optimization of Process Parameters

Niranjani Deshpande, Nihar Phalak and L.-S. Fan, The Ohio State University, Columbus, OH

Prior research has shown the effectiveness of carbon, derived from inexpensive sources like coal char, in controlling post-combustion NO_x emissions. The reduction of NO using char is a widely researched phenomenon. Some of the existing combustion modification methods such as re-burning employ this principle to reduce NO emissions. Pure carbon materials require very high temperatures for the C-NO reaction. However, the presence of mineral matter in char and the presence of O_2 in flue gas significantly reduce the required temperature of NO capture. Studies conducted at OSU have revealed that up to 98% reduction in NO levels using bituminous and lignite chars is achievable at moderate to high temperatures (500-800 °C) using simulated flue gas, at lab and pilot scale.

Naturally derived inexpensive calcium sorbent in the form of CaO or $Ca(OH)_2$ has been used extensively for the capture of CO2 and SO2. >90% CO_2 and complete SO_2 removal was achieved in studies conducted at OSU at the sub-pilot scale continuous reactor for the successful demonstration of the carbonation calcinations reaction (CCR) process.

In this study, these two reactive systems are combined to achieve single step CO_2, SO_2, and NO removal from gas mixtures at high temeperature (>500 °C). The effect of variables such as char particle size, use of activated carbon, extent of moisture in reactant gas, SO_2 concentration was quantified using a fluidized bed reactor. The fluidization behaviour of various mixtures of CaO sorbent and char was studied. The char selectivity for NO reduction was optimized. The results of these tests will be presented in the form of net NO, SO_2 and CO_2 capture. The merits and potential challenges of this novel process will also be discussed.

(739g) Calcium Looping Process for CO2 Capture - Investigation of Direct Carbonation of Ca(OH)2

Nihar Phalak, William Herold, Niranjani Deshpande and L.-S. Fan, The Ohio State University, Columbus, OH

CO_2 capture using calcium-based sorbents is rapidly developing into a viable technology. However, the decrease in CO_2 capture capacity of calcium sorbents when used over many cycles, presents a significant challenge. Among the various techniques that have been used to overcome this challenge, intermediate hydration seems most promising. Intermediate hydration results in the formation of $Ca(OH)_2$ making the traditional 2 step (CaO-$CaCO_3$) process a 3-step (CaO-$Ca(OH)_2$-$CaCO_3$) process. In such a scenario, the CO2 contacts the $Ca(OH)_2$ and not CaO. It has been believed that the $Ca(OH)_2$ decomposes to yield CaO which then reacts with CO_2. However, direct carbonation of $Ca(OH)_2$cannot be ruled out at the operating conditions.

The direct carbonation of $Ca(OH)_2$ was investigated in this work and the effects of CO_2 concentration and heating rate were studied. It was found that direct carbonation resulted in very high conversion to $CaCO_3$, as compared to indirect carbonation. It is believed that the conversion is enhanced due to equimolar counterdiffusion of H_2O which helps the CO_2 reach the unreacted CaO core through the $CaCO_3$ product layer. Direct carbonation also resulted in lower sintering due to the instant occupation of the pores formed because of H_2O evolution, by CO_2 (in the form of $CaCO_3$). This results in better carbonation conversion in the succeeding cycles. While CO_2 concentration did not affect direct carbonation conversion, different heating rates had an effect. Final conversions and reaction kinetics were poor when lower heating rates were used. Surface area and pore volume analyses were able to corroborate the theory of mitigation of sintering during direct carbonation. Using the shrinking core model (SCM), it was found that the effective diffusion coefficients during direct carbonation are higher than during indirect carbonation. The non-isothermal direct carbonation data was also fitted using the SCM and the experimental and model results were found to be in good agreement. These results will be presented during this talk and the implications on the process will be briefly highlighted.

Analysis of CO_2 Sorption/Desorption Kinetic Behaviors and Reaction Mechanisms with Steam on Li_4SiO_4

Zhang Qi, Han Daying, Liu Yang, Ye Qian, and Zhu Zibin
Dept. of Chemical Engineering, East China University of Science and Technology, Shanghai 200237, China

Introduction

The greenhouse effect and energy crisis are currently two of the most important problems worldwide. Hence, CO_2 capture technologies are required to decrease CO_2 emissions to the atmosphere and improve energy efficiency. They can be used in the direct separation of CO_2 from the high-temperature exhaust gases from power plants, particularly in in situ CO_2 sorption enhanced fuel steam reforming processes, which aims to produce pure hydrogen at lower temperature. The key point of a CO_2 sorption enhanced reaction system is to develop a satisfactory solid CO_2 captor with a large capacity, high selectivity, good cycle performance, and proper kinetic behaviors at relatively high temperatures. Thus far, several materials, such as hydrotalcite-like materials, CaO-based sorbents, and lithium-containing materials, have been proposed for CO_2 capture[1-4]. Among these materials, lithium orthosilicate (Li_4SiO_4) is considered to be one of the most potential materials. Li_4SiO_4 has been reported to adsorb CO_2 more than 30 times faster than Li_2ZrO_3 and is lighter and cheaper than Li_2ZrO_3. The mass uptake on Li_4SiO_4 due to CO_2 adsorption is almost 50% greater than the weight change for Li_2ZrO_3. It has also been used in sorption enhanced fuel steam reforming experiments, where it has shown an evidently promoting effect on the process.

Furthermore, in the methane steam reforming system, the existence of H_2O can not be ignored. Thus, the effect of water on the CO_2 sorption properties is very important for the application of the sorbents. Essaki and coworkers[5]observed a beneficial effect of adding water when absorbing CO_2 on Li_4SiO_4 at room temperature. Esther also investigated the CO_2 uptake profiles over Li_4SiO_4 and Li_2ZrO_3 at 525°C and 10% of CO_2 with and without water addition[6]. They also found the same trends of the positive impact of steam on the capture kinetics. These results make Li_4SiO_4 to be more practical in sorption enhanced steam reforming system. However, the reasons for the beneficial effect of steam addition on the CO_2 absorption kinetics of the CO_2 acceptors are not clear yet. These important phenomena need to be explained by kinetic mechanism.

Meanwhile, the investigation of the reaction system mechanism and design requires a suitable kinetic model for the CO_2 sorption/desorption process in Li_4SiO_4. However, only a few studies have reported relative models for Li_4SiO_4.

In this work,the CO_2 sorption/desorption kinetic behaviors on Li_4SiO_4 were analyzed. The theoretical compositions of the sorption/desorption reactions were calculated using FactSage. The sorption/desorption process on Li_4SiO_4 was investigated by comparing the shrinking core, double exponential, and Avrami–Erofeev models. Furthermore, the influence of steam on the CO_2 sorption process was theoretically analyzed. It has been observed that the presence of steam enhance the mobility of Li^+ and therefore the rate of diffusion control stage.

Results and Discussions

Theoretical Sorption/desorption behavior on Li_4SiO_4

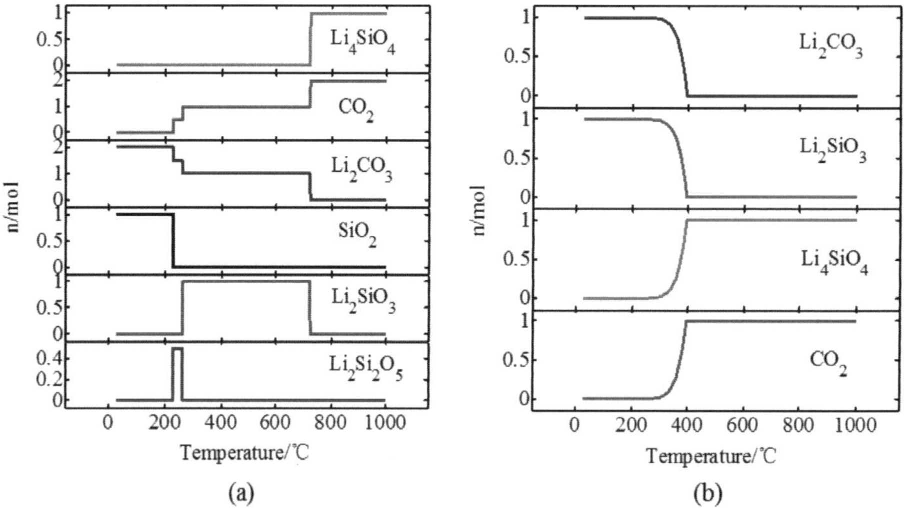

Figure 1. Thermodynamic equilibrium composition at different temperatures

(a) Sorption process. (b) Desorption process (initial condition: Li_2CO_3:Li_2SiO_3:N_2 = 1:1:1000)

Figure 1a shows the thermodynamic results of the CO_2 sorption process on Li_4SiO_4 under pure CO_2. When the temperature is higher than 724 °C, only Li_4SiO_4 and CO_2 exist, indicating a desorption process. Theoretically, Li_4SiO_4 can totally react with CO_2 in the first three stages with different products at different temperature ranges (Figure1a). Thus, the mass uptake of Li_4SiO_4 can be increased by improving the sorbent properties of Li_4SiO_4 when its conversion is lower than 100%.

Figure 1b shows the thermodynamic results of the CO_2 desorption process in pure N_2. The CO_2 desorption reaction equation is given as follows according to the results:

$$Li_2CO_3 + Li_2SiO_3 = Li_4SiO_4 + CO_2 \quad (1)$$

The CO_2 desorption reaction will not occur below 250 °C, and the reaction conversion increases with temperature at 250 °C to 395 °C. When the temperature is above 395 °C, the theoretical conversion of the desorption reaction is 100%. Thus, the CO_2 desorption temperature ranges from 395 °C to 1000 °C.

Effect of Steam on the CO_2 sorption Kinetics

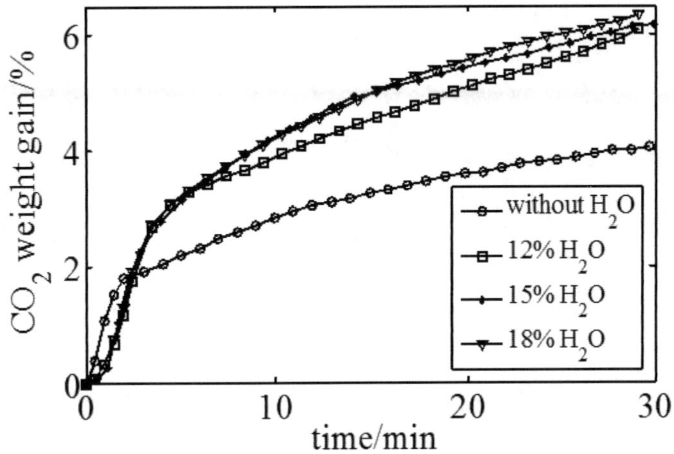

Figure 2. Change in mass uptake of Li_4SiO_4 over time at atmosphere with/without water at 575 °C and total pressure of 0.1MPa.

The influence of steam on the CO_2 sorption process was analyzed in the gas with H_2O concentration of 12% to 18% at 575 °C in a modified WRT-3P TG equipment. The result was show in Figure 2. It can be seen that with the addition of steam, the CO_2 mass uptakes increase 60% more than without steam after 30min adsorption. With the steam amount increasing, the adsorbent capture ability is enhanced accordingly. However, when the steam amount is greater than 15%, the effects of the steam amount on the sorption process is not obvious.

Table 1. Parameters of Avrami-Erofeev model for the CO_2 sorption process at 575 °C with/without water

water content /%	Inflexion point /min	rapid reaction stage			diffusion control stage		
		N	*K*	R2	*n*	*K*	R2
0	1	1.86	2.64E-03	0.9880	0.36	1.51E-06	0.9979
12	3	1.86	1.33E-03	0.9998	0.38	5.99E-06	0.9920
15	3	1.97	2.04E-03	0.9812	0.39	7.46E-06	0.9951
18	3	1.96	1.53E-03	0.9985	0.42	1.07E-05	0.9990

The sorption process with and without steam were analyzed using Avrami–Erofeev model. The kinetic parameters were summarized in Table 1. As it can be seen, the inflexion between rapid reaction stage and diffusion control stage is 1min in dry CO_2 atmosphere and increased to 3min by addition water. In the rapid reaction stage, the n value is all about 1.8~2, which indicates the CO_2 sorption reaction is controlled by the formation of the product crystals. The K value is 2.64E-3 in pure CO_2 atmosphere, and decreased firstly by steam addition. However, with the increase of steam amount, the reaction rate increases when the H_2O concentration is lower than 15%, and then decreases with H_2O concentration increasing. It is found that the effect of steam in the rapid reaction stage is not obvious. In the diffusion control stage, the n value and the reaction rate K value increase with the H_2O concentration. The reaction rate is 1.07E-5 when the water composition is 18%, which is 10 times larger than the value obtained in the dry CO_2 atmosphere, which means the reaction rate in the diffusion control stage enhanced with steam addition.

For the steam effects on the CO_2 sorption kinetics can be summarized as follows:
Hydrolysis:

$$CO_2 + H_2O \leftrightarrow HCO_3^- + H^+ \leftrightarrow CO_3^{2-} + 2H^+ \quad (2)$$

$$Li_2O + H_2O \leftrightarrow 2Li^+ + 2OH^- \quad (3)$$

Reaction:

$$2Li^+ + CO_3^{2-} \leftrightarrow Li_2CO_3 \quad (4)$$

$$2H^+ + 2OH^- \leftrightarrow 2H_2O \quad (5)$$

Total reaction:

$$Li_2O + CO_2 \leftrightarrow Li_2CO_3 \quad (6)$$

The presence of steam is believed to enhance the mobility of Li^+ and therefore the rate of the reactions, which is the reason that the steam effect on the diffusion control stage is more obvious than the rapid reaction stage and make the adsorption rate in diffusion control stage increased with the H_2O concentration. On the other hand, the formed Li_2CO_3 shell, which is on the surface of the adsorbent particle, is slightly dissolved in water with the solubility of 1.31g Li_2CO_3 /100g H_2O and then the absorption is also promoted[7]. Thus, the inflexion point increased from 1min to 3min.

Kinetic analysis of CO_2 desorption process

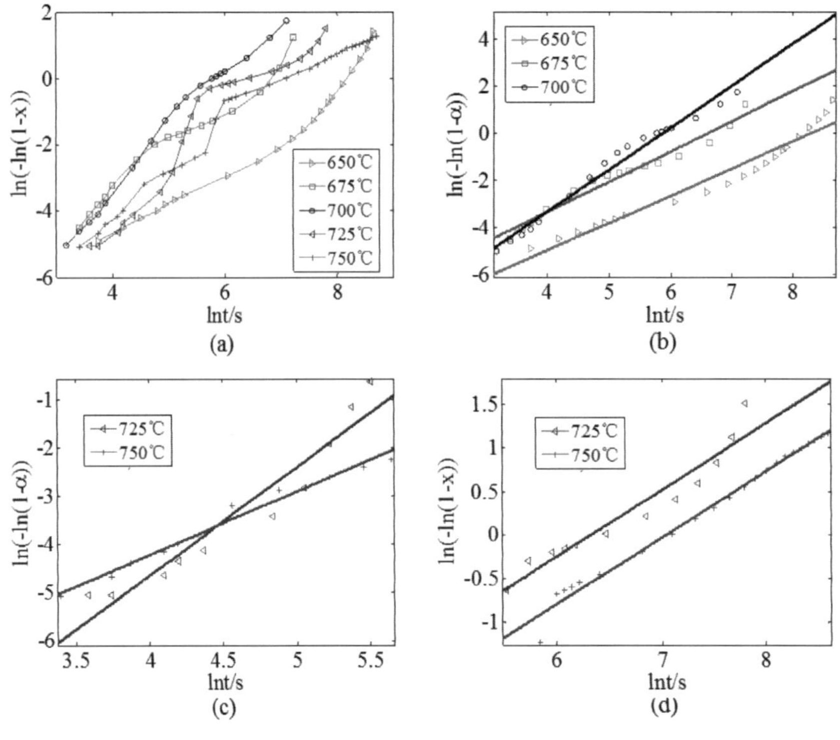

Figure3. Fit of the CO_2 desorption kinetic experimental data with the (a) Avrami–Erofeev equation in the (b, c) first and (d) second stages

The Avrami–Erofeev model was also used to analyze the CO_2 desorption process. Figure 3a shows the ln(-ln(1-x)) versus lnt line of the experimental data obtained in the CO_2 desorption process. The curves are nearly straight below 700 °C, but at 725 °C and 750 °C, the two processes have inflection points of approximately 5 min, as shown by the slope. The two curves are split into two stages at 5 min to analyze the kinetic behavior in detail. Figures 3b, 3c, and 3d show the linear fit at different temperatures, and the n and K values were obtained according to the fitting results.

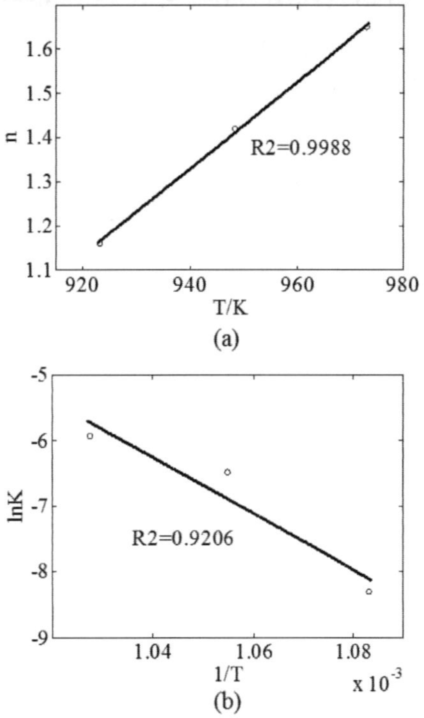

Figure 4. Kinetic parameters of the Avrami–Erofeev equation for the desorption process [(a) n versus T line; (b) ln K versus $1/T$ line]

Figure4a shows the linear fit of n as a function of temperature. Below 700 °C, the slopes are all greater than 1 and increase with the temperature. This result means that the whole desorption process is controlled by the reaction rate of the formation and growth of the product crystals. The temperature dependence of n when the temperature is below 700 °C is given by Eq. (7).

$$n = 0.0098T - 7.8639 \tag{7}$$

Figure 4b shows the fitting of the K values with the Arrhenius equation when the temperature is below 700 °C and the active energy is 3.57 E+5 J/mol, which is approximately 13 times larger than that of the rapid reaction stage in the CO_2 sorption process. As shown in the figure, the formation and growth of the product crystals is more difficult in the desorption process than in the sorption process.

When the temperature reaches 725 °C, the slope n in the first stage has values of 2.25 and 1.32 for 725 °C and 750 °C, respectively. These values indicate that the CO_2 desorption process is controlled by the formation and growth of the product crystals. In the second stage, both n values decrease to 0.76, indicating that the reaction proceeds under the diffusion control. This phenomenon is caused by the liquefaction of Li_2CO_3 considering that its melting point is 723 °C. The liquefied Li_2CO_3 comes in full contact with Li_2SiO_3, the process is first controlled by the rate of the formation and growth of the product crystals. With the CO_2 formation, the gas assembles on the surface of Li_2SiO_3 because it hardly diffuses through the liquid film and the contact area of the reactants decreases. Thus, in the second stage, the CO_2 desorption process will proceed under the diffusion control.

The Avrami–Erofeev model is also suitable for the CO_2 desorption process. A desorption temperature near 700 °C should be chosen to maintain a relatively high reaction rate. This kinetic model can be used to simulate the CO_2 sorption enhanced fuel steam reforming process as well when the desorption temperature is lower than 723 °C (the melting point of Li_2CO_3).

Conclusion

This work aims to develop a suitable kinetic model for investigating the kinetic behaviors and reaction mechanism of the CO_2 sorption/desorption process on Li_4SiO_4.

The Avrami–Erofeev model was found to be suitable for the CO_2 sorption process on Li_4SiO_4 within a wide temperature range. It can describe the rapid reaction stage, which is neglected in the other two models. Moreover, the reaction mechanism can be explained using the Avrami–Erofeev model combined with the double-shell mechanism, and the Avrami–Erofeev equation is also suitable for analyzing the kinetic behavior of the CO_2 desorption process.

The presence of steam is believed to enhance the mobility of Li^+ and therefore the rate of the reactions, therefore, the effects of steam on the rapid reaction stage is not obvious, whereas the reaction rate of the diffusion control stage are 10 times larger with 18% H_2O concentration than the value obtained in the dry CO_2 atmosphere, which indicates the possibility of CO_2 capture technology with Li_4SiO_4 for the steam existence system.

For the desorption process, the temperature should be near 700 °C to maintain a relatively high reaction rate. During the process, Li_2CO_3 reacts with Li_2SiO_3 on the interface to form a dense shell in the middle covering the unreacted Li_2SiO_3. Then the CO_2 desorption process proceeds with the diffusion of Li^+ and O^{2-} through the solid Li_4SiO_4 shell and CO_2 through the Li_2CO_3 shell. The whole desorption process is controlled by the rate of the formation and growth of the Li_4SiO_4 crystals.

Acknowledge

This work was financially supported by the National Natural Science Foundation of China (Grant No.21176080), the Natural Science Foundation of Shanghai (Grant No.09ZR1407600).

•

References

Use “References” as a heading name, with your listing that follows in the base font size (12pt). Number each reference consecutively (1, 2, 3, . . . etc.) and single space between each reference. Indent after each reference number. Do NOT use end notes. Refer to the example below.

1. Reynolds PS (2006), “Carbon dioxide capture from flue gas by pressure swing adsorption at high temperature using a K-promoted HTlc: Effects of mass transfer on the process performance,” *Environmental Progress*, 25, pp. 334-342.
2. Hufton JR, Mayorga S, Sircar S(1999), “Sorption-enhanced reaction process for hydrogen production,” *AIChE Journa,* 45, pp. 248-256.
3. Koumpouras G, Alpay E, Lapkin a, Ding Y, Stepanek F(2007), “The effect of adsorbent characteristics on the performance of a continuous sorption-enhanced steam methane reforming process,” *Chemical Engineering Science*, 62,pp. 5632-5637.
4. Lysikov A, Trukhan S, Okunev A(2008), “Sorption enhanced hydrocarbons reforming for fuel cell powered generators,” *International Journal of Hydrogen Energy*, 33, pp. 3061-3066.

5. Essaki K, Nakagawa K, Kato M, Uemoto H(2004), "CO2 absorption by lithium silicate at room temperature," *Chem. Eng. Jpn*, 37, pp. 772–777.
6. Ochoa-Fernández E, Zhao TJ, Rønning M, Chen D(2009), "Effects of Steam Addition on the Properties of High Temperature Ceramic CO2 Acceptors," Journal of Environmental Engineering, 135,pp. 397–403.
7. Essaki K, Kato M, Nakagawa K, Uemoto H(2004), "CO2 Adsorption by Lithium Silicate at Room Temperature," *Journal of Chemical Engineering of Japan*, 37, pp. 772-777.

(751a) Simulation of Lithium-Ion Cells and Batteries

Ralph E. White, Chemical Engineering, University of South Carolina, Columbia, SC

Simulation of lithium-ion cells and batteries includes predicting the performance and life of these devices. The beginning of life performance of lithium-ion cells can be predicted by using John Newman's porous electrode theory, which will be reviewed in this presentation. Predicting the rate of heat generation in lithium-ion cells is a critical aspect of designing heat removal components of a lithium-ion battery pack for electrical vehicle applications. The rate of heat generation of the components in a lithium ion cell can be predicted by using multiscale, multidimensional modeling technique, which will be presented in this talk. In order the reduce the computation burden for solving these model equations, reduced order models have been developed and will be outlined in this tutorial. The life of lithium-ion cells and batteries can be predicted by adding equations that describe the degradation processes to the model equations, or by using parameter estimation techniques to determine how the important physical parameters in the beginning of life model equations change with time. An overview of these life predicting techniques will be presented.

(751b) Design Aspects of Flow Batteries for Large Scale Energy Storage

Robert Savinell, Chemical Engineering, Case Western Reserve University, Cleveland, OH

Large scale energy storage technology is needed to facilitate integrating renewable energy generation technologies, such as wind and solar, into the electric grid because these energy generation technologies are inherently intermittent and thus matching supply to load demand is a challenge. Also, energy storage is needed for modernizing the grid into a "smart" electricity distribution system. There is a large variety of potential energy storage solutions available for this broad range of applications and understanding the various operating and economic constraints and requirements is necessary to choose a technology. Flow batteries of various chemistries are gaining attention again in recent years because of these markets and opportunities and of the inherent advantage of flow batteries. The advantages and disadvantages, and their technical and economic challenges will be addressed for a number of approaches. The presentation will focus its attention of electrode materials, electrode design, and cell design for redox flow batteries.

(751c) Electrode and Cell Design Principles in Batteries and Fuel Cells

Trung V. Nguyen, University of Kansas, Lawrence, KS

In a battery or fuel cell, the electrodes and other components in the cell must be carefully selected and designed to address the electrode potentials, reactions, and transport (electronic, ionic, molecular) processes, especially the limiting ones, of the system to meet the performance (discharge and charge rate and cycle life), shelf-life or storage life, and safety requirements. While high active surface area is needed for high power, its effect on the self-discharge rate and corrosion should also be considered. In addition to the active materials that participate in the electrochemical reactions, the role of the supporting materials such as the current collectors, catalyst supports, electrolytes, and binders, both conductive and non-conductive, must also be considered because they can have a significant effect on the performance and/or shelf-life of the system. This presentation will discuss the materials selection and components and cell design requirements for a few battery and fuel cell systems.

(751d) Electrowinning Principles and Practice: A Brief Introduction to the Electrochemistry and Engineering Behind Modern Electrowinning Plants

Dan Steingart, Energy Institute and Chemical Engineering, City College of New York, New York, NY

Electrowinning, or electroextraction, is the oldest industrially applied electrochemical process, and the backbone of the production of many metals, both precious and otherwise (aluminum, copper, zinc and lead, among others), as well as industrial gases (e.g. the chlor akali process). In the US alone electrowinning processes consume on the order of 1 to 2% of total electricity produced, and many of the most energy intensive processes are have much room for improvement of efficiency. In this tutorial we will broadly examine the theory and practice of both the electrochemical processes as well as the supporting mining and extraction operations. Beyond a discussion of the basic operating principles, we will examine issues of feed purity, flow control, and rate sensitivity. Finally, experimental and R&D scale approaches to efficiency improvements will be covered. While the aluminum, copper, and chlor-alkali systems will be systems of focus in this talk, the principles discussed will be applicable to any modern electrowinning process.

(751e) Design of Enzyme Electrodes for Sensing and Power Applications

Joshua Gallaway, Chemical Engineering, Energy Institute, City College of New York, New York, NY

The prospect of harnessing the catalytic functionality of enzymes for practical applications holds tremendous scientific and technological promise. Composed of folded proteins built around an active site, enzymes are biological catalysts that enable thousands of highly specific reactions to occur under mild conditions. In biological systems, electron exchange to and from enzymes happens diffusionally from small redox molecules. In contrast, practical systems designed by humankind often rely on electron exchange at a heterogeneous interface (an electrode). By using mediator molecules to transport electrons to enzymes they can be used on electrodes with relatively high currents, but the resulting systems can be complex. Electron hopping, Michaelis-Menton kinetics, materials aspects of hydrogels, and other complications make enzyme electrode design less than intuitive. This talk aims to demonstrate the connection between electrode modeling, synthesis techniques, and fabrication, which when used together can greatly improve the practical performance of enzyme electrodes. An example case will be presented in which current response of an oxygen reducing enzyme electrode was doubled by molecular engineering.

(774a) Development of Proton Selective Membranes for Efficient and Economical Vanadium Redox Flow Batteries

Justin Walls[1], Pedram Jahanian[2], Guangzhao Mao[2], Dr. Steven Salley[3] and Dr.Simon Ng[3], (1)Chemical Engineering, Wayne State University, Detroit, MI, (2)Chemical Engineering and Materials Science, Wayne State University, Detroit, MI, (3)Chemical Engg., Wayne State University, Detroit, MI

Development of Proton Selective Membranes for Efficient and Economical Vanadium Redox Flow Batteries

Justin Walls, Padram Jahanian, Guangzhao Mao, Steven Salley, K. Y. Simon Ng

Department of chemical Engineering and Materials Science, College of Engineering,

Wayne State University, Detroit MI 48202

Flow Redox Batteries (FRBs), especially the Vanadium Redox Flow Battery (VRFB), provide a promising chemistry for storing intermittent renewable energy on a very large scale. While these batteries have already been developed on the megawatt hour level; none of the FRB technologies has seen significant market penetration. This is mainly because the technology is still very expensive, due to high material costs, and less than satisfactory performance parameters and reliability.

The key component of the VRFB, which is one of the leading causes of the elevated cost and unreliability, is the ion exchange membrane (IEM). The membranes need to possess good ion conductivity while maintaining high mechanical and chemical stability. Perfluorosulfonic polymers such as Nafion are the most common membranes for proton exchange in the VRFB due to their good chemical stability and high proton conductivity. However, due to the tendency of vanadium to pass through the membrane, Nafion membranes suffer from decreased energy efficiency. This project is focused on decreasing the vanadium ion diffusion through the membrane by modifying the nanostructure of Nafion. There has been extensive research activity in the modification of Nafion membranes, for example, modification by layer-by-layer polyelectrolyte multilayer films and in situ sol-gel reaction to close the pores by silica nano particles. Our project combines these two methods using hyper-branched polyelectrolyte multilayer films to further lower the vanadium ion diffusion rate. Nafion modification by the sol-gel chemistry was conducted by treating the Nafion in the tetraethylorthosilicate (TEOS) methanol solution. Poly(diallyldimethylammonium chloride) (PDDA) and poly(sodium styrene sulfonate) was used in the layer-by-layer deposition. Diffusivity and ion crossover of these membranes was tested and compared to that of the unmodified membranes as well as other modification techniques that have been reported. A small scale VRFB system was constructed to test the performance of these modified membranes, and efficiency and reliability of the system as a whole was measured.

(774b) Microporous Separators On Fe/V Redox Flow Battery: A Valuable Opportunity for Cost Reduction

Xiaoliang Wei[1], Qingtao Luo[1], Bin Li[1], Zimin Nie[1], Yuyan Shao[1], Feng Chen[1], Baowei Chen[1], Guanguang Xia[1], Liyu Li[2], Zhenguo Yang[2] and Wei Wang[1], (1)Pacific Northwest National Laboratory, Richland, WA, (2)UniEnergy Technologies, LLC, Mukilteo, WA

Microporous Separators on Fe/V Redox Flow Battery: A Valuable Opportunity for Cost Reduction

Wei, Xiaoliang; Luo, Qingtao; Li, Bin; Nie, Zimin; Shao, Yuyan; Chen, Feng; Chen, Baowei; Xia, Gordon; Li, Liyu; Yang, Z Gary; Wang, Wei

Pacific Northwest National Laboratory, 902 Battelle Boulevard, PO Box 999, Richland, WA 99354

Redox flow batteries are considered as one of the most promising medium-to-large scale energy storage technologies and have attracted much attention both academically and industrially.[1],[2] A number of redox chemistries have been proposed, and significant progress has been achieved by today.[3],[4] However, broad market penetration of redox flow batteries is still hindered by their intrinsic limitations such as high cost and low durability of components, narrow operational temperature range, limited solubility of active species, and/or low electrochemical activity.[5] The recently invented iron-vanadium (Fe/V) redox flow battery employing $Fe^{2+/3+} - V^{2+/3+}$ redox couples have shown to be a promising option for stationary energy storage.[6],[7] The Fe/V flow battery demonstrated stable cycling performance with a nearly 100% utilization ratio over a broad temperature range of 0-50°C.

The positive charged species, Fe^{3+}, is a relatively weak oxidant. Therefore, hydrocarbon-based ion exchange membranes and/or separators are possible options for use in Fe/V flow battery system. This contribution investigated the cycling performance of a variety of polyethylene microporous separators on Fe/V flow cell. Among them, some separators exhibited energy efficiency of around 70% at temperatures ranging from 5-50°C and at current densities up to 80mA/cm^2. Because these separators are very inexpensive, their use significantly reduces the capital cost of Fe/V flow battery, delivering great potential for developing a low-cost energy storage system.

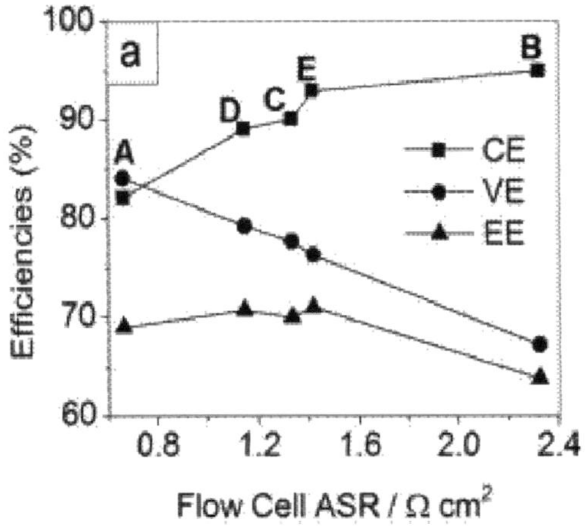

Figure 1 Flow cell cycling efficiencies (CE, VE, and EE) of microporous separators A, B, C, D, and E.

[1] Z. Yang, J. Zhang, M. C. W. Kintner-Meyer, X. Lu, D. Choi, J. P. Lemmon, and J. Liu, *Chem. Rev.* 2011, *111*, 3577-3613.

[2] J. Rugolo, and M. J. Aziz *Energy Environ. Sci.* 2012, **DOI:** 10.1039/C2EE02542F.

[3] M. Skyllas-Kazacos, M. H. Chakrabarti, S. A. Hajimolana, F. S. Mjalli, and M. Saleem *J. Electrochem. Soc.* 2011, *158*, R55-R79.

[4] L. Li, S. Kim, W. Wang, M. Vijayakumar, Z. Nie, B. Chen, J. Zhang, G. Xia, J. Hu, G. Graff, J. Liu, and Z. Yang *Adv. Energy Mater.* 2011, *1*, 394–400.

[5] S. Eckroad, *Vanadium Redox Flow Batteries: An In-Depth Analysis.* EPRI, Palo Alto, CA: 2007. 1014836.

[6] W. Wang, S. Kim, B. Chen, Z. Nie, J. Zhang, G. Xia, L. Li, and Z. Yang *Energy Environ. Sci.* 2011, *4*, 4068-4073.

[7] W. Wang, Z. Nie, B. Chen, F. Chen, Q. Luo, X. Wei, G. Xia, M. Skyllas-Kazacos, L. Li, and Z. Yang *Adv. Energy Mater*. 2012, *2*, 487–493.

(774c) In Situ Investigation of Vanadium Ions Transport in Redox Flow Battery

Qingtao Luo[1], Liyu Li[2], Zimin Nie[3], Wei Wang[1], Xiaoliang Wei[1], Feng Chen[1], Bin Li[4], Guanguang Xia[1], Baowei Chen[5], Gary Yang[1] and Sprenkle Vincent[1], (1)Pacific Northwest National Laboratory, Richland, WA, (2)Hydrocarbon Processing Group, Pacific Northwest National Laboratory, Richland, WA, (3)Pacific Northwest National Labratory, Richland, WA, (4)Pacific Northwest National Laboratory, Richland , WA, (5)Biological Sciences, Pacific Northwest National Laboratory, Richland, WA

The all-vanadium redox flow battery (VRB) is probably the most promising and extensively researched flow battery system. In order to shed light on the VRB capacity fading mechanism, it is necessary to investigate the transport behaviors of vanadium ions across the membrane. In this work, flow batteries with vanadium and iron redox couples as the electroactive species were employed to investigate the transport behavior of vanadium ions in the presence of an electric field. It was shown that the transport of vanadium ions was affected by a combination of different factors, such as the charge/discharge process, current density, and SOC. The electric field accelerated the positive-to-negative and reduced the negative-to-positive transport of vanadium ions in the charging process and affected the vanadium ion transport in the opposite way during discharge. With the increase of current density, the vanadium ion transport rate increases during the charge process but decreases during discharge. At each given current density, the vanadium ion transport rate, in general, demonstrated a trend of decreasing with the increase of SOC.

(774d) Electrolyte Optimization for Fe/V Redox Flow Battery System

Bin Li[1], Wei Wang[1], Zimin Nie[2], Vincent Sprenkle[1], Baowei Chen[3], Xiaoliang Wei[1], Qingtao Luo[1], Guanguang Xia[1], Liyu Li[4] and Zhenguo Yang[4], (1)Pacific Northwest National Laboratory, Richland, WA, (2)Pacific Northwest National Labratory, Richland, WA, (3)Biological Sciences, Pacific Northwest National Laboratory, Richland, WA, (4)UniEnergy Technologies, LLC, Mukilteo, WA

Electrolyte optimization for Fe/V redox flow battery system

Bin Li, Wei Wang, Zimin Nie, Vince Sprenkle, Baowei Chen, Xiaoliang Wei, Qingtao Luo, Guanguang Xia, Liyu Li and Zhenguo Yang

Pacific Northwest National Laboratory

902 Battelle Boulevard, PO Box 999, Richland, WA 99354

Abstract

Redox flow batteries (RFBs) research and development have recently enjoyed a renaissance primarily due to their ability to store large amounts of power and energy, up to multi-MW and multi-MWh, respectively, for renewable energy integration or smart-grid deployment.[1] Broad market penetration, however, has not been achieved due to several limitations such as high cost, instability in long-term operation, and narrow operational temperature range, etc. Recently invented Fe/V redox flow battery (IVRFB) has attracted more and more attentions due to its long-term cycling stability and the freedom to utilize membranes other than expensive Nafion.[2] In this study, extensive matrix study was performed on the Fe/V electrolyte in order to determine the relationships of the composition, state of charge (SOC), and temperature stability. As we know, the temperature stability of the RFB systems is of paramount importance for practical use. A RFB system with wider temperature stability will not only enable the RFB to be deployed with less geographical limitation, but also possibly eliminate the active temperature management system reducing parasitic loss. As previously reported, the current Fe/V electrolyte cannot be operated or stored at lower temperatures ($\leq 0^oC$) due to the precipitations ($FeCl_2$). In this research, factors influencing the stability of electrolytes in both positive and negative half-cells with different SOCs at different temperatures from -10^oC to 40^oC were systematically studied. Through the optimization of the electrolyte composition, an electrolyte stable in the temperature range from -5^oC to 40^oC was identified. Electrolyte stability data and IVRFB flow cell cycling performances using low-cost separators will be presented (Fig.1).

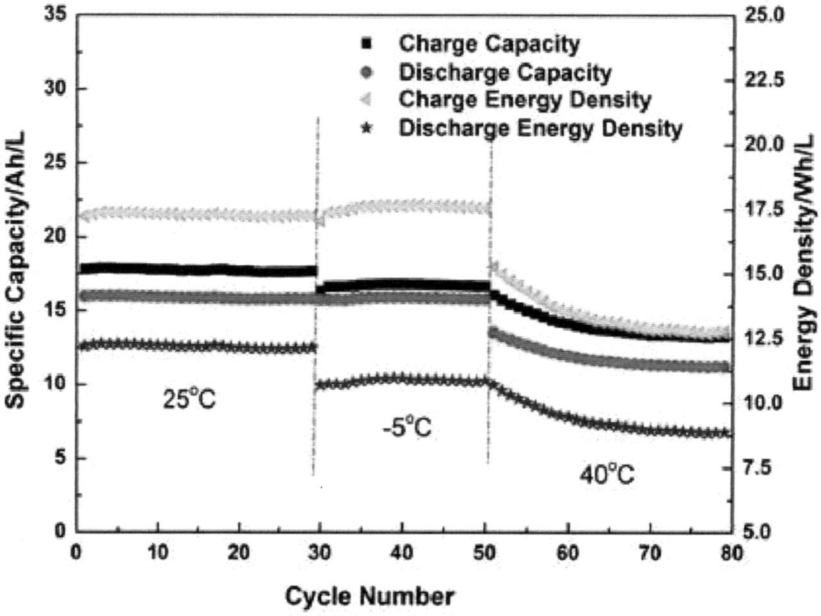

Fig.1 Cycling performances of optimized Fe/V electrolytes

References

(1) Yang, Z.; Zhang, J.; Kintner-Meyer, M. C. W.; Lu, X.; Choi, D.; Lemmon, J. P.; Liu, J. *Chemical Reviews* **2011**, *111*, 3577.

(2) Wang, W.; Kim, S.; Chen, B.; Nie, Z.; Zhang, J.; Xia, G.-G.; Li, L.; Yang, Z. *Energy & Environmental Science* **2011**, *4*, 4068.

(774e) Electrostatically Assembled Carbon Nanoparticle Electrodes for Vanadium Redox Flow Batteries

Abhinandh Sankar and **Anastasios Angelopoulos**, School of Energy, Environmental, Biological, and Medical Engineering, University of Cincinnati, Cincinnati, OH

Efficient charge and mass transfer at the surface of the electrode is critical to overall redox flow battery (RFB) performance. Fundamental structure-property correlations for RFB electrodes cannot presently be obtained using electrodes produced from carbon felts or carbon nanoparticle slurries and pastes due to the ill-defined distribution of active sites in these materials. For example, a recent study using high surface area multi-wall carbon nanotubes as electrode materials yielded only a two to three-fold increase in peak current density relative to glassy carbon. We here report on the first use of electrostatic layer by layer (LbL) assembly to fabricate RFB electrodes. Since its introduction by Iler over four decades ago, LbL assembly has been repeatedly shown capable of producing well-defined arrangements of nanoparticles and the method is now ubiquitous in scientific research. However, the approach has yet to be applied to produce well-defined electrode structures for RFB electrodes and has only rarely been explored in the general field of electrocatalysis. Key areas of concern include: 1) the impact of the polymer counter-ion layer that is typically required for assembly on ohmic resistance, 2) accessibility of active sites in tightly packed layered structures, and 3) durability. In this investigation, we demonstrate how LbL assembly of graphite nanoplatelets may be used to achieve unprecedented high current densities for the all-vanadium system through fine control over the thickness and morphology of the electrode. In particular, we show how nanoparticle packing density, size, and size distribution in addition to local polymer binder concentration and crystalline order all impact active site concentration and promote reagent accessibility to these sites in a manner that has not been previously recognized. Such basic understanding permits rational optimization of electrode chemistry and structure using carbon-based electrocatalysts as well as detailed interpretation of the promotional effects of inorganic electrode additives. Finally, we demonstrate through electrochemical cycling how these coatings are exceptionally stable relative to electrodes fabricated without LbL assembly.

(774f) Fractional Degradation Cost for Cycling a Zinc/Bromine Flow Cell Battery

Alex Bistrika[1], Alexandre F.T. Yokochi[1], Michael Antonishen[2], James Davidson[2], Douglas Halamay[2], David Naviaux[3], Jiajia Song[2], Ted Brekken[3] and Annette von Jouanne[3], (1)School of Chemical, Biological and Environmental Engineering, Oregon State University, Corvallis, OR, (2)Electrical Engineering and Computer Science, Oregon State University, Corvallis, OR, (3)School of Electrical Engineering and Computer Science, Oregon State University, Corvallis, OR

It is widely understood that intermittent resources such as solar and wind are dependent on environmental factors outside our ability to control. This allows for the possibility of either power output exceeding or falling short of forecast levels which leads to more frequent cycling of regional hydro generators and greater overall grid instability under the current infrastructure. Quantifying the cost associated with added maintenance attributed to an increased penetration of wind and the physical limitations of a growing dependence on Pumped Hydro Energy Storage (PHES), becomes essential in painting a more complete picture. However, alternative means of Large Scale Energy Storage (LSES), for storing excess power when the outputs exceeds demand and to supplement output power when it falls short of demand, is a widely accepted route for mitigating intermittent power generation.[1] Thus, it is crucial to establish an accurate and relative cost function for the commercially available and most economically viable storage technologies. The most promising electrical energy storage technology, to replace PHES, is flow cell batteries.[2] This begs the question what are the degradation mechanisms in these building size chemical systems and what do we do when they stop working. In this study we report on the key degradation mechanisms that could be expected for standard operation of a zinc/bromine flow cell battery using graphite electrodes and assign a cost function dependent of system operating parameters. In a previous study we reported the oxidation (or charging) reaction as the most damaging to the electrode material, see **Figure 1**. We discuss aging trends and electrode performance by considering surface chemistry as observed by Raman spectroscopy and scanning electron microscopy. Theoretical system performance based on steady state current density measurements is used to establish a normalized operating cost function for a zinc/bromine battery, as can be seen in **Figure 2**.

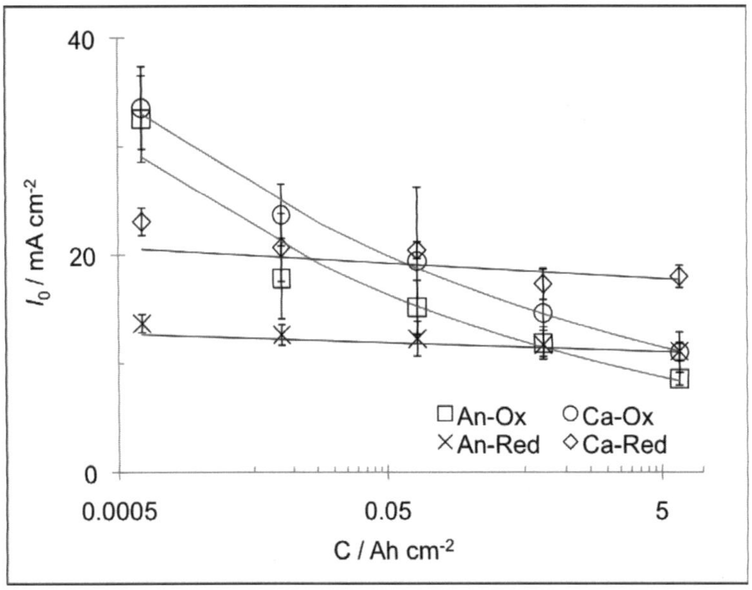

Figure 1 Oxidation and reduction exchange current; measured for both the cathode and anode in 6 [M] NH4Br. The x-axis represents the electrode aging, depicted by total coulombs passed through the electrode per unit surface.

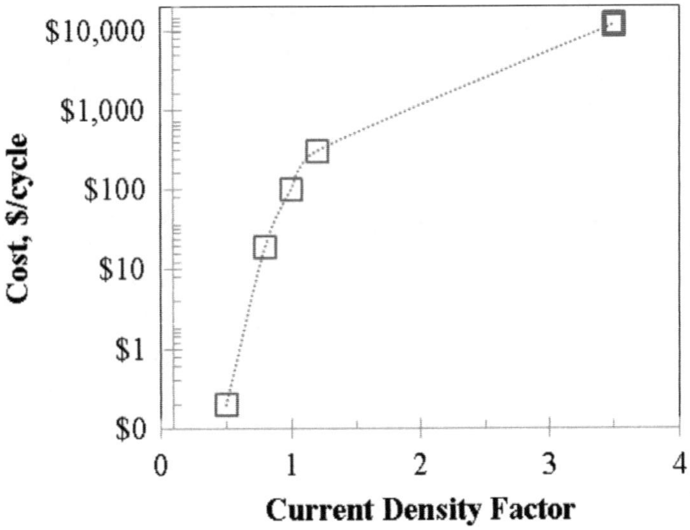

Figure 2 Cost of a 10 % state of charge cycle with respect to a representative operating parameter (current density factor).

1. Yang Z, Zhang J, Kintner-Meyer MCW, et al. Electrochemical Energy Storage for Green Grid. *Chemical Reviews.* 2012/04/07 **2011**;111(5):3577-3613.

2. Soloveichik GL. Battery Technologies for Large-Scale Stationary Energy Storage. *Annual Review of Chemical and Biomolecular Engineering.* 2012/04/24 **2011**;2(1):503-527.

(774g) Design Mesostructured LiMn2O4 As Low-Cost Cathode Materials for Lithium Ion Batteries

Jian Zhu, Chemical Engineering, Wayne State University, Detroit, MI, K. Y. Simon Ng, Department of Chemical Engineering and Materials Science, Wayne State University, Detroit, MI and Da Deng, Chemical Engineering and Materials Science, Wayne State University, Detroit, MI

Lithium ion batteries are the dominant power sources for mobile electronic devices and will play increasingly important role in expanding market of electric vehicles and green energy storage. Spinel $LiMn_2O_4$, alternative to expensive and toxic $LiCoO_2$, as cathode materials in lithium ion batteries are attracting much attention for advantages of low cost, environmental benignness, and abundance. However, the low lithium-ion diffusion coefficient and capacity fading caused by dismutation reaction of Mn^{3+} restrict the applications of $LiMn_2O_4$ at large/industrial scale where bulky batteries are required to have good cycling performance at high current rate with no significant capacity fading. Our strategy is to design and tailor the structure and composition of $LiMn_2O_4$ to improve its conductivity and cycling performance. In this presentation, our latest efforts on the design and facile fabrication of $LiMn_2O_4$ as high-power cathode materials will be reported. The unique structure is tailored and controlled to allow fast lithium ion diffusion while maintaining structure stability. The unique mesostructure is fully characterized by electron microscopy and x-ray diffraction and their electrochemical performance in lithium ion batteries is thoroughly evaluated.

(791a) Nanomaterials Enabled Lithium Chemistry for Advanced Rechargeable Batteries

Chao Wang, Materials Science Division, Argonne National Laboratory, Argonne, IL

Employment of nanomaterials in electrode assembly is promising for the development of advanced lithium-ion batteries. Nanostructured electrodes can potentially achiever higher power and better cyclability than their bulk counterparts, due to the facilitated charge transfer at electrode/electrolyte interface, electronic/ionic transport in the electrodes, and accommodation of strain during lithium insertion/extraction by reduction of material dimensions. Moreover, design of electrodes based on nanomaterials can enable new lithium insertion/extraction reactions that are not possible with bulk materials, and thereby expand our scope of material choices for energy storage. These advantages have lead to substantial literature effort on the synthesis of electrochemically active nanomaterials for battery applications.

Here we report the development of novel manganese oxide nanoparticles (10 - 100 nm) by organic solution synthesis and their solid chemical conversion into lithium manganese oxide ($Li_xMn_2O_4$) spinels for cathode application in lithium ion batteries. We managed to demonstrate that by reducing the material dimensions to nanoscale, charge/discharge up to $0<x<2$ can be achieved which however is not possible for conventional bulk or microscale materials (with only $0<x<1$). Our work shows the promise of expanding the scope of material choices for energy storage by size scaling and the great of potential of improving battery performnace by employing advanced nanomaterials.

(791b) Dynamics of Pumped Heat Energy Storage Systems

Hugo S. Caram and Fan Ni, Chemical Engineering, Lehigh University, Bethlehem, PA

The eventual success and widespread utilization of photovoltaic and wind energy generation will depend on the availability of efficient and economical storage capability. The current technology for long term (diurnal), high efficiency large scale energy storage of electrical energy appears to have been limited to Pumped Hydroelectric Storage (PHS) and Compressed Air Storage (CAES). Both of them are dependent of the availability of geological features such as elevated lakes or empty deep mines for PHS or caverns for the CAES. This features are not available everywhere and other technologies such flow batteries while capable of large scale storage are limited in their power output.

Pumped heat energy storage systems can provide a valuable alternative PHS or CAES. The proposed process uses a Brayton cycle where during the loading cycle heat is pumped to heat up a high pressure thermal regenerator while a low pressure thermal regenerator is cooled down to a low temperature. During the unloading period the gas circulation is reversed. Cold gas from the low pressure regenerator is compressed and heated in the high pressure regenerator to be then expanded to the low pressure vessel. Electrical energy storage capacities in the 30-50kWh/m^3 range appear possible and are comparable to lead acid batteries (25-50 kWh/m^3).

Analysis and optimization of these systems are complicated by the difficult dynamics of thermal regenerators. A novel matrix algorithm is proposed that allows us to directly calculate cyclic steady state as well as transients. The effects of different design parameters such as pressure ratios, compressor and expander efficiencies, type of gases, solid heat capacities and gas solid heat transfer coefficients are explored. Possible synergistic integration with natural gas combustion combined cycles will be discussed.

The effective utilization System for Renewable Energy Making use of Regional Characteristic

Masahiro Yamamoto, Tsuguhiko Nakagawa
Graduate school of *Okayama Prefectural University,*
Kuboki 111, Soja-shi, 719-1197, Okayama, Japan

ABSTRACT

The SMART factory project has been developing a process to reduce CO_2 emissions through researching and developing a process to make plastic from biomass at the biomass accumulated factory in Maniwa, Okayama.

At the biomass accumulated factory, wood chips have been made from shavings and unused wood parts from sawmills. The biomass factory consists of a building chips plant which makes the wood chips, a building silo which stores chips, and a barkyard which crushes bark. The energy used at the biomass accumulated factory consists of electric power and liquid fossil fuels. Liquid fossil fuels have been used by forklift trucks and others vehicles for wood chips. The CO_2 emissions by liquid fossil fuels are accounted for about 43% of the total from the biomass accumulated factory. In order to reduce 25% CO_2 emissions, the SMART factory is planned to be supplied by renewable energy from the wood chips plant and photovoltaic power generation.

In this study, in order to achieve this aim, the mathematical simulation model which calculates the energy demand and supply per hour in the SMART factory has been developed. Based on the mathematical simulation model, a system which is able to reduce consumption of liquid fossil fuels and electric power has been designed.

As a result, it has been proved that it is possible to reduce 25% or more CO_2 emissions using the system that combines photovoltaic power generation with electric forklift trucks.

1. INTRODUCTION

The SMART factory project in Maniwa has been developing a process to reduce CO_2 emissions using renewable energy through researching and developing a process to make plastic from biomass at the biomass accumulated factory.

In this study, we have examined the system which substituted fossil fuels to be used as fuel for mobile equipment such as forklift trucks for renewable energy such as photovoltaic power generation (hereafter refered to as "PV") which will be constructed at the biomass accumulated factory.

2. OUTLINE OF THE NEW ENERGY SYSTEM

Outline of the new energy system is shown in Fig.1. At the biomass accumulated factory, mobile equipment such as two forklift trucks, shovel loaders and two grapples have been used. These vehicles have previously used liquid fossil fuels. Among this mobile equipments we considered that forklift trucks be substituted for electric forklift trucks and the system that charges the PV electricity to the battery of electric forklift trucks.

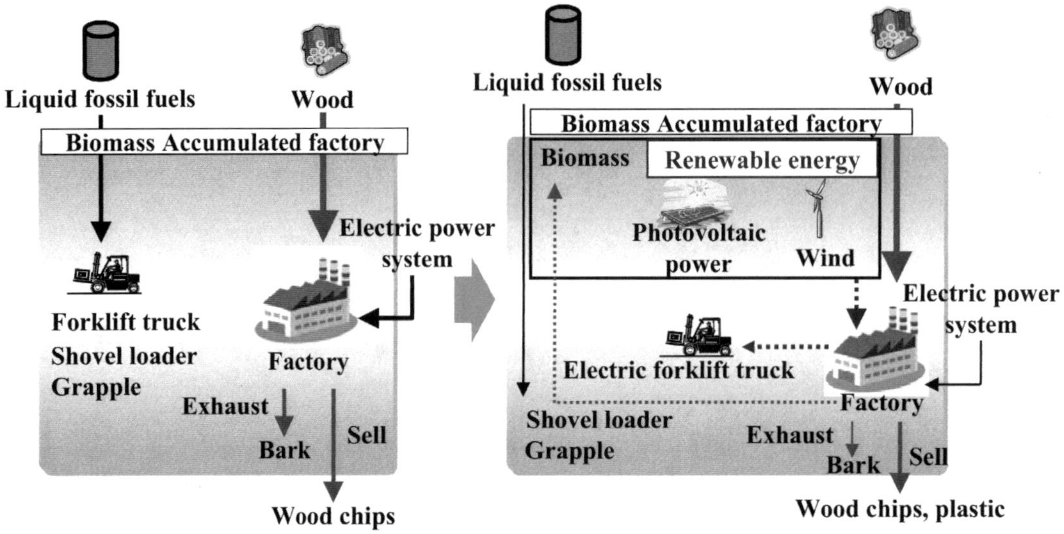

Fig.1. Images of the new energy system

3. THE EXAMINATION METHOD

3.1 Model Equation

For each production of wood chips, using equation (1), the mathematical simulation model which is able to calculate the energy demand and supply per hour in a year (8760hr) has been designed.*[1]

$$\sum_{i=1}^{n} \alpha_i = \frac{X_i}{\eta} + E_i - \beta_i \quad \cdots \quad (1)$$

where α_i : electric power from electric power system [kWh]

E_i : electric power demand of factory [kWh]

X_i : electric power demand of electric forklift truck [kWh]

β_i : electric power generation of PV [kWh]

η : Charging&discharging efficiency (=0.85)

3.2 Production of wood chips and energy consumption

The relationship of monthly production of wood chips and electric power demand, liquid fossil fuels consumption of forklift truck at the biomass accumulated factory is shown in Fig.2 and Fig.3.

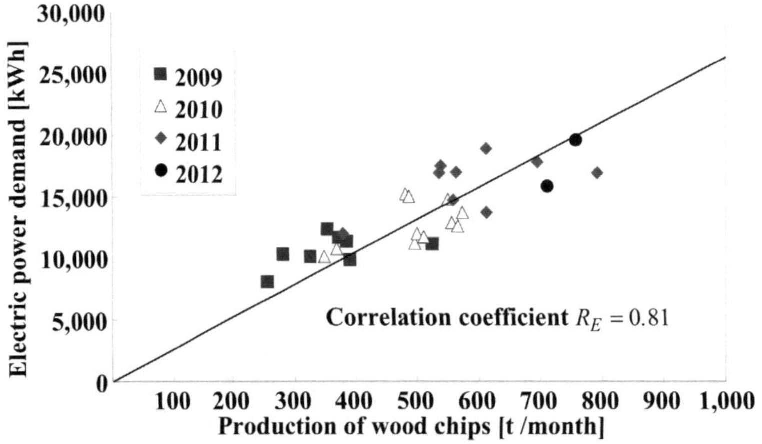

Fig.2. Relationship between electric power demand and production of wood chips

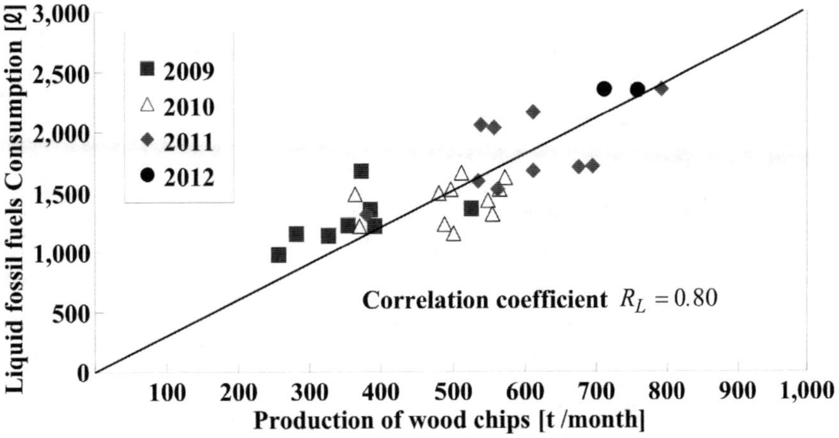

Fig.3. Relationship between liquid fossil fuels consumption of forklift trucks and production of wood chips

In Fig.2, it can be seen that there is a correlation between electric power demand and production of wood chips. For least-squares method,

$$y = 26.3x \quad \cdots \quad (2)$$

From equation (2), it can be estimated electric power demand from production of wood chips. On the other hand, in Fig.3, similarly, it can be obtained by equation (3).

$$Z = 0.9x \quad \cdots \quad (3)$$

where x : Production of wood chips [t/month]

y : electric power demand [kWh]

Z : liquid fossil fuels consumption of forklift trucks [ℓ]

Using equations (2) and (3) in this new energy system the optimal operation performance can be calculated depending on the production level of wood chips.

3.3 Result of examination

Regarding production of wood chips at 400[t/month] and 700[t/month], the relationship of CO_2 emissions and area of PV at the biomass accumulated factory is shown in Fig.4.

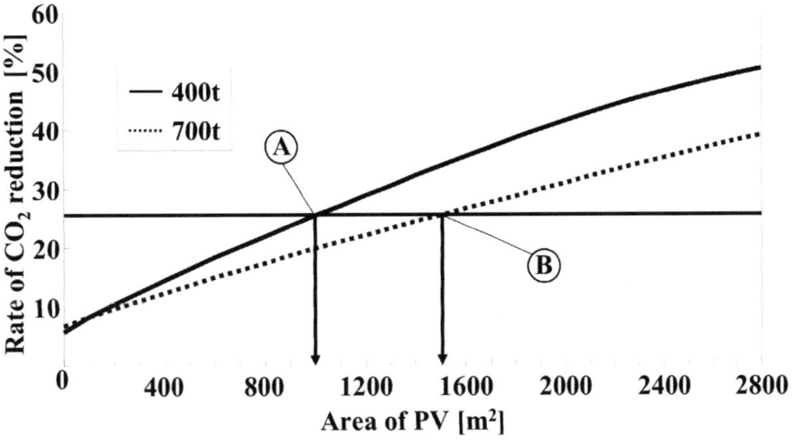

Fig.4.Relationship area of PV and rate of CO_2 reduction per production of wood chips

In Fig.4Ⓐ,Ⓑ, the breakdown in the case of reducing 25% CO_2 emissions is shown Fig.5 Ⓐ,Ⓑ. In Fig.4, PV is of low electric power generation efficiency. But, it is able to be installed on a sloping site e.g. the factory roofs. Also, it is amorphous that installation costs are low. Maximum output per unit area is 69.1[W/m^2].

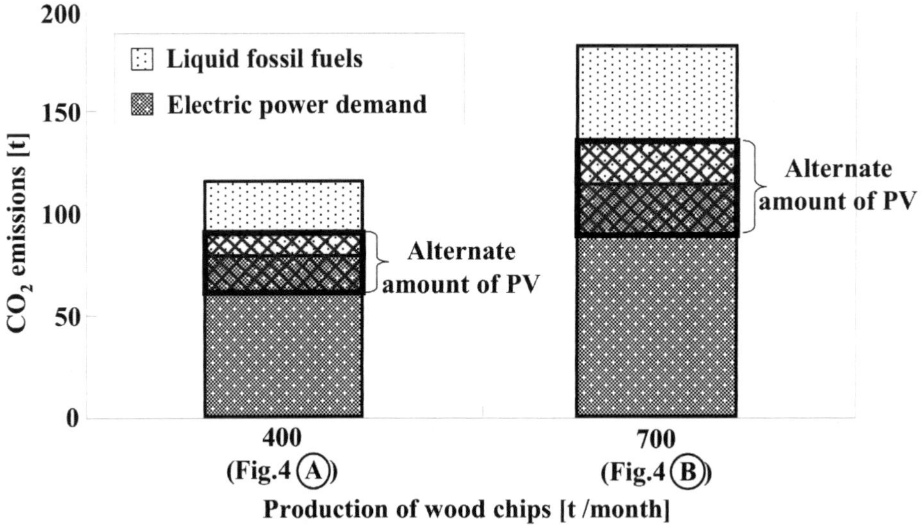

Fig.5. The breakdown in the case of reducing 25% CO_2 emissions

As a result, when production of wood chips is 700[t/month], it has been proved that the PV installation area should be greater than 1500m^2 to reduce 25% or more CO_2 emissions at the biomass accumulated factory.

4. CONCLUSION

Depending on the rate of Production of wood chips, it has been proved that it is possible to estimate liquid fossil fuels and electric power demand and calculate the design area of PV depending on the target of CO_2 emissions reduction.

In this study, we are subsidized by the funds form the Overall Promotion of Ministry of Education, Culture, Sports, Science and Technology. I am deeply grateful to them for their supports.

Reference

1. Shimohara, K., Y. Notoji, T. Nakagawa "Distributed Energy System Combination with Energy Conservation, PV and EV", *Kagaku kogaku Ronbunshu* Vol.38 No.4, pp.255—262, Japan (2012)

(791d) Thermochemical Cycle of a Mixed Metal Oxide for Augmentation of Thermal Energy Storage in Solid Particles

Brian D. Ehrhart[1,2], Nathan P. Siegel[3], Eric N. Coker[2] and Alan W. Weimer[1], (1)Chemical and Biological Engineering, University of Colorado at Boulder, Boulder, CO, (2)Sandia National Laboratories, Albuquerque, NM, (3)Mechanical Engineering, Bucknell University, Lewisburg, PA

Solid particle heat transfer and storage media have been shown to be able to operate at temperatures >1000°C in concentrated solar power (CSP) applications, much higher than the operational limit of 600°C for currently state-of-the-art molten nitrate salt heat transfer fluid. Solid particles can be endothermically reduced by direct exposure to concentrated solar energy, thus absorbing and storing thermal energy beyond that possible with sensible heating alone. The particles can then be oxidized exothermically at a later time, releasing the stored chemical heat and effectively augmenting the thermal energy storage capacity of the solid particles. A mixed metal oxide spinel material that reduces in the temperature range of interest (1000°C-1200°C) has been examined for applicability to this thermochemical energy storage concept. A description of this application, prospective materials, and details of the thermochemical cycle are presented. The heats of reduction and oxidation for the thermochemical cycle have been determined for various operating conditions to evaluate the amount of thermal energy that may be stored. Additionally, the extent of reduction possible at various operating conditions has been measured.

(791e) Carbohydrate Could Be the Best Electricity Storage Compound and a High-Density Hydrogen Carrier

Y.-H. Percival Zhang, Biological Systems Engineering, Virginia Polytechnic Institute and State University, Blacksburg, VA and Chun You, Biological Systems Engineering, Virginia Polytechnic Institute and State University

Electricity is a high-quality secondary energy, which can be generated from renewable primary energies and fossil fuels. Carbon-neutral electricity generation is becoming more and more important, but most of them, such as wind and solar electricity, suffer from large output variations. Therefore, it is vital to store electricity in the form of chemical compounds. Among numerous potential chemical compounds, such as hydrogen, isobutanol, methanol and hydrocarbons, we suggest that carbohydrates (e.g., starch, cellulose, glucose) could be the best electricity storage compound with a combustion energy density of 17 MJ/kg and an electricity output density of >3,000 Wh/kg [1]. We designed a novel biological CO2 fixation pathway that can fix CO2 by using hydrogen or electricity for producing carbohydrate mediated by cell-free cascade enzymes [1]. In this talk, we will present our efforts in developing building blocks and building modules for this ambitious project and conduct proof-of-concept experiments. Through the above bioconversion plus our previous achievements such as theoretical conversion of carbohydrate to hydrogen or electricity [2] and low-cost enzymatic conversion of cellulose to starch (unpublished), the electricity-carbohydrate-hydrogen cycle would not only supplement current and future primary energy utilization systems for facilitating electricity and hydrogen storage and enhancing secondary energy conversion efficiencies, but also address such sustainability challenges as transportation fuel production, CO_2 utilization, fresh water conservation, food and feed production, and maintenance of a small closed ecosystem in emergency situations [1]. In a word, carbohydrate would become a new oil in the future.

1. Zhang Y-HP, Huang W-D: **Constructing the electricity-carbohydrate-hydrogen cycle for a sustainability revolution**. *Trends Biotechnol.* 2012, **Epub. DOI: 10.1016/j.tibtech.2012.02.006**.

2. Zhang Y-HP, Evans BR, Mielenz JR, Hopkins RC, Adams MWW: **High-yield hydrogen production from starch and water by a synthetic enzymatic pathway**. *PLoS One* 2007, **2**:e456.

(791f) Energy Storage Design for Electric Transmission Systems

Donald J. Chmielewski, Illinois Institute of Technology, Chicago, IL

It is widely recognized that a major concern with the installation of renewable energy systems is the fact that wind and solar sources are non-dispatchable. That is, the power produced from renewable systems is dependent on environmental conditions and is likely uncorrelated with the power demand from load centers. In contrast, fossil based sources typically have the ability to track demand, but depending on the technology possess limits with respect to response time. Thus, many have advocated the use of massive energy storage systems to capture energy produced from renewable sources and then dispatch as load centers demand. An additional concern is the fact that renewable sources tend to be located far from the load centers (again in contrast to most fossil sources which are typically placed near demand). This combination of spatial and temporal mismatch with load center demand has caused many to worry about the stability of the electric power grid. Clearly, wide swings in power throughput over great distances will cause the weakest parts of the grid to fail.

To address both the spatial and temporal issues surrounding renewable power generation we have developed a scheme aimed at analyzing the capabilities of a regional power generation and distribution system. In particular, the scheme proposes a supervisory control law aimed at utilizing a mix of renewable and fossil sources along with energy storage systems to meet power load demands. This proposed control law takes into consideration not only the power output capacity of each source but also the rate at which power output can change as well as the storage capacity and efficiency losses associated with the storage units. In addition, the proposed scheme incorporates the maximum and minimum throughput limits on the distribution lines. If time allows, we will present an extension of the scheme aimed at sizing and placing energy storage systems as well as locating new power distribution lines.

AUTHOR INDEX

AUTHOR INDEX

AUTHOR INDEX

AUTHOR INDEX

American Institute of Chemical Engineers
3 Park Avenue
New York, New York 10016-5991

CURRAN ASSOCIATES INC
proceedings.com